OECD DOCUMENTS

MINING AND NON-FERROUS METALS POLICIES OF OECD COUNTRIES

PUBLISHER'S NOTE

The following texts have been left in their original form to permit faster distribution at lower cost.

ORGANISATION FOR ECONOMIC CO-OPERATION AND DEVELOPMENT

ORGANISATION FOR ECONOMIC CO-OPERATION AND DEVELOPMENT

Pursuant to Article 1 of the Convention signed in Paris on 14th December 1960, and which came into force on 30th September 1961, the Organisation for Economic Co-operation and Development (OECD) shall promote policies designed:

— to achieve the highest sustainable economic growth and employment and a rising standard of living in Member countries, while maintaining financial stability, and thus to contribute to the development of the world economy;
— to contribute to sound economic expansion in Member as well as non-member countries in the process of economic development; and
— to contribute to the expansion of world trade on a multilateral, non-discriminatory basis in accordance with international obligations.

The original Member countries of the OECD are Austria, Belgium, Canada, Denmark, France, Germany, Greece, Iceland, Ireland, Italy, Luxembourg, the Netherlands, Norway, Portugal, Spain, Sweden, Switzerland, Turkey, the United Kingdom and the United States. The following countries became Members subsequently through accession at the dates indicated hereafter: Japan (28th April 1964), Finland (28th January 1969), Australia (7th June 1971) and New Zealand (29th May 1973). The Commission of the European Communities takes part in the work of the OECD (Article 13 of the OECD Convention).

Publié en français sous le titre :

EXTRACTION ET MÉTAUX NON FERREUX :
LES POLITIQUES DES PAYS DE L'OCDE

FOREWORD

In order to enhance the transparency of national policies in the mining and non-ferrous metals sector, the OECD High Level Group on Commodities requested the OECD Secretariat to undertake a comprehensive review of relevant Member country policies. It was decided that the review would cover the main producing and consuming Member countries. As a guide, countries were chosen which account for more than 3 per cent of the western world production or consumption of any one of the six main non-ferrous metals. Sixteen reviews were completed covering fifteen Member countries and the European Community. The exercise, in the view of the High Level Group on Commodities, did make a contribution to furthering transparency in the sector and the Group recommended that the results be derestricted.

This study is the result of the national reviews. They have been structured on the basis of a standard format to provide for consistency in the analytical approach and to facilitate comparison. The reviews outline the main policies affecting the industry, from exploration and mining through to metals production, as well as horizontal policies which have an impact on the industry. Policies relating to metal consumption and supply were also covered. In addition to the national reviews, an overview in the first chapter synthesises the main features of the relevant national policies.

This study has been prepared under the supervision of Dale Andrew and Denis Audet of the Secretariat and benefited from comments of national authorities. Nonetheless the report is published on the sole responsibility of the Secretary-General of the OECD.

CONTENTS

Mining and processing facilities

Post-Tokyo Round average tariffs on non-ferrous metals

Tariff treatment according to stage of processing of six non-ferrous metals

European Community

OVERVIEW

PROSPECTING AND EXPLORATION

Granting of exploration permits

Prospecting and exploration rights are closely related to the principle of law governing ownership of sub-surface, i.e. the mineral rights. Each country surveyed has adopted one of two fundamentally different principles of law. Mineral rights are governed either by (a) the principle of separation of surface and underground property rights, with the state owning the sub-surface and hence the mineral rights, regardless of who owns the surface or (b) the principle of non-separation according to which underground rights are held by the surface landowner, whether a private owner or the state.

The first of these, and the most widely found regime among OECD Members, is referred to as the regalian regime. The regime of non-separation, referred to as the common law tradition, prevails in the United States, the United Kingdom and Ireland[1].

This difference in legal regime governing mineral rights has a direct bearing on the role of the state in the allocation and exploitation of such rights and in devising procedures for granting prospecting permits and mining concessions. Conditions attached to the transfer or allocation of underground rights for commercial purposes vary widely among countries which follow the same principle of law. Differences in administrative organisation and conditions attached to the granting of rights also have economic implications which may effect the economic efficiency of exploration and mining activities.

In countries following the common law tradition where there is non-separation of surface and underground rights, private landowners determine the conditions for the transfer of their rights for exploration or mining purposes against payment, usually in the forms of royalties. In these cases, the role of the state is limited to the application of horizontal laws (environment, taxation, safety, etc.). However, this principle of non-separation has been eroded by several land reforms in Ireland and sale practices in the United Kingdom, whereby many owners gave up surface lands while retaining mineral rights. The determination of underground rights in these two countries can be a complex exercise. Nevertheless, the state assumes its ownership rights over public lands by stipulating the conditions for the granting of mineral rights for exploration and mining purposes. For example, in the United States, where a significant portion of its territory is not owned privately, exclusive ownership rights over public lands are granted to prospectors at no cost, against commitments of minimal assessment works. In turn these rights give access to mining permits if it is demonstrated that an economically viable deposit exists.

In countries where the sub-surface rights are vested in the state, each state has devised a regulatory framework for the granting of mining rights according to its specific natural endowment and policy preferences for securing mineral needs from domestic sources. Where surface land is owned privately, prospectors must reach an agreement with the surface owner to carry out exploration. As an exception to that rule, in France and Greece, the state can proceed without consent of private landowners to carry out exploration activities. An exploration right usually gives access to an exclusive mining right. However, many countries may exercise a veto over the transfer of exploration and mining rights to third parties. Similarly, the transfer of exploration and mining rights to foreigners is subject to special authorisations in many countries. For example, in Greece, granting exploration rights to foreign individuals or companies is subject to approval by the Council of Ministers.

Lands availability

Competition amongst various land uses, such as recreation, natural parks, animal and bird sanctuaries, native land claims and land conservation, has increasingly resulted in restricting access for exploration and mining activities. Restrictions may be either explicit or implicit as the conditions attached for obtaining exploration rights over restricted areas may be so cumbersome that in practice exploration activities are no longer carried out. An exceptional case appears to be France where an interministerial agreement lays down that prospecting activities throughout the national territory, including reserves and natural parks, cannot be banned provided that an environmental impact study has been conducted beforehand.

In the United States, although land access is free in principle, there are increasing restrictions de facto which limit access. Since the 1960s restrictive regulations issued by the Forest Service Lands and the Bureau of Land Management have made access to public domains considerably more complex by markedly limiting or banning exploration and prospection activities over a considerable area of the United States. Over 30 per cent of United States land is owned by the federal government, of which 90 per cent lies in the 12 western states including over 40 per cent in Alaska alone.

Access for exploration and mining may come into conflict with rights over native lands. In Australia, the Federation and several states have recognised the rights of aboriginals to own the land on which they live and hence the right to grant or turn down prospecting permits (except where this would be in the national interest), as well as to negotiate compensation directly for damage to their land resulting from prospecting or mining activities. However other states argue that the land has been borrowed from the aboriginals and that the granting of exploration permits remains under their control. In Canada, the granting of exploration permits and mining claims is a matter for the government department responsible for mines in the region in question. Through the native land claims process, natives in Canada are acquiring fee simple title to significant areas of land, particularly in northern Canada. This land is not legally alienated from mineral production inasmuch as it might be in the interest of natives to develop resources in their territory. In this case, firms wishing to explore or develop on these lands must negotiate access conditions with the native landowners.

For regalian tradition countries that apply the principle of separation of surface and sub-surface, restrictions applied to exploration because of specific land uses may amount to defeating the principle of separation in that underground uses become linked to prescribed surface uses. On the other hand, it may be claimed that in these cases the state willingly decides to exercise its ownership rights by forbidding commercial operations in the sub-surface beneath prescribed surface areas. The mining industry has expressed concern that such increasing restrictions on land access for exploration purposes are jeopardising the long term availability of minerals and metals. This tendency appears both in regalian tradition countries, as well as for public lands in common law countries.

Geological services

The countries surveyed generally have a specialised agency responsible for carrying out basic geological surveys and producing geological maps of the national territory. These agencies gather and provide geoscientific knowledge about the national territory, including offshore areas. Some provide geological engineering expertise and logistic assistance for scientific activities. Often they are also the depository of the findings of geological prospecting by companies. They therefore collate a vast amount of scientific and technical information on which subsequent exploration may be based.

In view of their level of skill and considerable specialisation, these services are well placed to be consulted in policy formulation related to domestic exploration, environmental regulations and occasionally on exploration technologies. Several of these agencies are involved in the assessment of the geologic potential of territories outside their borders. For example, in the United States, France, Germany, Spain and the United Kingdom, the relevant agency undertakes geological prospecting operations abroad, often in co-operation with the national geological services.

In countries with a federal structure such as Canada or Australia, federal bodies are responsible for collecting data and geological exploration in close co-operation with state/provincial agencies under joint programmes. It should also be mentioned that the EC has recently acknowledged the importance of closer co-operation amongst its Member States to facilitate access to and the exchange of geological data.

Financial and tax incentives

The financial risk associated with exploration activities is relatively high given the low probability of discovering mineral deposits of commercial value. To stimulate exploration activities through other means than direct government participation, governments have recourse to various forms of financial and tax incentives. For example, grants repayable in the case of commercial finds are provided in Japan, Italy, Ireland, and Spain. Various forms of tax incentives, including the treatment of exploration expenditures as capital expenses, are available in many countries. However, recent taxation reforms, which have lowered tax rates and widened the base of taxable activities, have reduced sector-specific tax incentives, including for this activity. Thus in Canada and Ireland, this resulted in the removal of specific tax schemes for exploration purposes. In Canada, until 1990, tax deductions were offered for earned depletion in conjunction with a system, known as "flow-through shares", enabling exploration firms to transfer unused exploration tax deductions to investors. In general, the mineral industry has expressed concern that the absence or the removal of tax incentives will divert investment away from exploration, thus adversely affecting the discovery rate of non-ferrous deposits. Table 1.1 summarises the measures applied in the various countries.

Direct participation

Whereas all countries surveyed provided some indirect support generally through the taxation system, only a few countries have direct government participation in exploration. Governments of several continental European countries and Japan are involved in exploration activities through direct participation of state agencies or in providing grants to subcontractors for fulfilling prescribed policy objectives and projects. For example, in France, Italy and Greece, the state participates directly in exploration efforts through state or para-statal agencies. In Australia, the respective Department of Mines for each state/territory confines its actions to preliminary exploration and geological mapping. In two other regalian countries, the state does participate actively, but favours co-operation programmes. In Japan, since 1966 the government has been operating and funding basic geological prospecting programmes implemented in co-operation with scientific bodies, private firms and municipalities. In Sweden, the government agency responsible for prospection activities has a budget representing approximately one-third of the total investment in exploration in the country. In countries with a dominantly common law background, exploration activities are confined to private initiatives with financial support limited to tax incentives. In the United States, two exploration programmes had featured direct government participation, but the last of these was terminated in 1974.

MINING AND METALS PRODUCTION

Direct government intervention

Analysis of the OECD countries' policies reveals that direct government intervention in mining and metals production is absent not only in the United States and United Kingdom, two common law countries, but also in several regalian countries: Germany, Japan, Australia and Canada. In Ireland, state participation in the biggest mine in that country ceased in 1989. For the remaining countries the principal types of direct intervention in the non-ferrous metals sector identified are:

Nationalisation

This consists in extensive nationalisation of the mining sector through policy and regulatory action enabling the government to control parent companies and banks or other main shareholders. Such is the case in France, where the state has been involved at all stages in the metals production cycle, including first and second processing stages since 1981. Again in Portugal, government intervention in this sector expanded markedly after the 1974 revolution.

Direct government aid to ailing firms

Direct aid covers direct capital or similar assistance (any type of participatory loan), either within the legal framework of nationalised firms with the shareholder increasing own resources or through programmes of aid to ailing firms. This tool has been extensively used in Spain, Italy and Greece. Usually the impetus for such aid originated in the oil shocks of 1973 and 1979-80.

In Spain, the National Industry Institute (INI) has been helping firms in difficulty since 1980, especially in the aluminium and copper processing sectors. In Italy, the government controls almost all the non-ferrous metals industry because during the 1980s when companies were rescued and financially restructured, the government gradually took over the sector through direct capital subsidisation. In Greece, after 1982 there was direct government participation in mining and metals production, largely resulting from the conversion of company debts into risk capital by government bodies. The government intervened through a specialised agency, nationalised banks or by taking over the management of ailing firms.

Direct participation in mining development

In other countries, the fact that the government reserves the right to intervene in mining projects can also be regarded as a form of direct aid to mining development. An example is the 51 per cent holding of a Portuguese government agency (EDM) in the Neves Corvo mine or the acquisition by the Irish Government of a 25 per cent holding in Tara Mines. In Sweden, the state holds a pre-emptive right to 50 per cent for any mining project, but this right has apparently always been relinquished.

Indirect government intervention

Governments in the countries surveyed also provide aid to this sector indirectly: through various financing mechanisms, the taxation system, state-controlled power utilities or regional development programmes. As a rule, countries have a package of financial and taxation support measures, consisting of reduced interest rate loans, state guarantees for loans from private bodies, subsidies, etc.

Financial assistance

Financing through subsidised interest rates with long grace periods and state guarantees for loans contracted by firms has been an important instrument of indirect government aid to the non-ferrous metals industry, albeit recently in decline. Investment in energy savings, environmental control, R&D, staff training, etc. have been common targets of such government financial assistance which the non-ferrous industry has made use of in recent years.

In the United States and Canada, there are few financing programmes specific to this sector although firms may be entitled to obtain funds for staff training, R&D or environmental control investment programmes. In Australia, the commonwealth and states may extend subsidised loans to the non-ferrous metals industry, as well as guarantees on loans from private, national or international banks.

During the 1970s, the Japanese Government set up the Emergency Loan Fund. Administered by the Mining and Metals Agency of Japan (MMAJ), the Fund supports mining producers in difficulty. The state guarantees MMAJ loans from commercial banks and subsidises the difference between market interest rates and those applied to mining firms. Other subsidised financing programmes concern energy-saving investments. Under these programmes, together with temporary tariff-quotas, the Japanese Government guided the scale-down of the Japanese aluminium industry from an annual capacity of 1.6 million tonnes per year (tpy) in 1978 to about 350 000 tpy in the late 1980s.

In the EC, structural funds are available to Member States, including: the European Regional Development Fund (ERDF), the European Investment Bank (EIB) and the European Social Fund (ESF). These bodies finance projects in less-favoured regions, develop new sources of production and assist in the closure of mines, although non-ferrous mining and metals projects have not been priorities for structural fund interventions.

Taxation

Taxation has traditionally been a potent instrument of indirect government intervention in the sector. It should be noted that practically all taxation systems were reformed during the 1980s, with many of the taxation incentives abolished (accelerated depreciation of investments, the establishment of special tax allowances, etc.). Removal of these incentives has been offset by considerable reductions in the general taxation rates applied to company earnings (from between 15 to 20 percentage points). While it may be argued that these reforms have markedly increased transparency in the mining sector, they have thereby removed the specificity or priority nature which had traditionally been granted to this sector in many countries. The sector now more or less falls within the general rules of corporation taxation. Table 1.2 summarises the main taxation systems currently in force in the countries surveyed.

Table 1.2 reveals major differences in tax systems in EC Member States and the Community authorities have recognised the need for more homogeneous and favourable taxation arrangements for the development of mining activities. The following aspects are on the EC agenda: depletion allowances; possibility of offsetting losses incurred during the first few years of activity against future profits and tax exemptions for environmental control investments.

Inputs pricing

Governments also affect competitive advantage for their non-ferrous metal industries through tariffs or pricing policy relating to inputs, particularly for transport and energy. Electricity costs are a significant cost component of non-ferrous smelting operations, particularly for copper and zinc smelting and aluminium production. Thus the economic viability of non-ferrous smelting and refining operations is dependent on securing reliable low cost electricity supplies. Since electricity production is usually a state monopoly, governments frequently intervene in the fixing of electricity rates.

The general consensus is that the principle of any healthy pricing policy should consist in making large consumers pay tariffs which take into account at least part of the long-term marginal and social costs (e.g. representing the cost of long-term replacement of energy production and distribution facilities). Pricing should therefore be based on marginal costs rising in the long term. In practice, energy producers usually set tariffs according to stock levels, in other words, the water level in reservoirs and equipment capacity utilisation. In periods of surplus production or where high energy consuming customers are liable to be lost, energy producers are prepared to negotiate short-term discounts on normal tariffs.

Successive oil shocks in the 1970s led to a shift in comparative advantage for non-hydrocarbon-based power generation, particularly towards hydro and nuclear power. Investment in new non-ferrous smelting capacity tended to locate near these cheap energy sources. After initial surplus capacity in new power plants, increased demand brought an end to the energy glut around 1985 and energy producers began to offer flexible tariffs indexed to metal prices. Examples of flexible indexed tariffs in place include those between the Bonneville Power Administration (BPA), a public company, and aluminium producers in the Northwest United States[2], between Hydro-Québec and large consumers which settled in Quebec after 1987[3] and between Alcoa and the Victoria state Government in Australia for the Portland aluminium works[4]. In 1991, the Norwegian Parliament allowed Norwegian firms to opt either for flexible indexed tariffs or a tariff rising in real terms[5].

Flexible indexed tariffs seem to work to the mutual benefit of electric utility companies and aluminium producers. As an example, BPA, which relies heavily on aluminium smelters for its revenue, has designed its flexible tariff rate to maintain stable aluminium smelter loads and thereby enhance BPA's revenue recovery from these customers over the long run. Simultaneously, the smelters can maintain optimal utilisation rates over longer periods of time. Some flexible indexed tariff agreements, such as the one between Hydro-Québec and aluminium smelters in that region include an actual profit-sharing component, while others, like BPA's, do not.

Since the end of the 1980s, Elecrictrité de France (EDF), the French state electricity monopoly, has also had to cope with a large power surplus, again due to an over ambitious investment programme in nuclear energy and slower growth in demand than forecast. While increasing exports and cutting back its equipment plans, EDF has supported the establishment of industries with a high energy consumption in France through its pricing policy. EDF has given preferential rates to Péchiney for its aluminium smelters[6].

Regional development programmes

Countries reviewed, including the EC, operate regional development programmes to address economic disparities among regions. Generally, these regional development programmes are targeted at assisting a balanced development and the mining and non-ferrous metals sector are among the potential candidates for assistance. state assistance is generally available to all industries by means of direct subsidies, guaranteed loans, capital grants and tax relief. Few examples were found where regional development programmes were a major means of assistance to the non-ferrous sector. The Aughinish alumina refinery benefited from special programmes for attracting investment and helping to develop less favoured regions of Ireland. In Canada, mineral development programmes are used, along with similar programmes for other sectors, as a means of meeting the federal government's regional development objectives. In general, they make information in the areas of geoscience, mining and mineral technology research and market studies and promotion widely available to the industry.

HORIZONTAL POLICIES

Foreign investment controls

Mineral resources are a sector for which restrictions on foreign ownership have proliferated. This is often a matter of national security in view of the resource's role in the national economy or the strategic value associated with some minerals. Introduced in past periods of greater government control on the economy, restrictions affecting parts of the processing chain from ores to metals have often not been removed. Some of these restrictions are explicit and transparent but others are implicit. The granting of mineral rights to foreigners is restricted in many countries by the non-transferability of rights without state approval. Mineral rights for specific commodities are sometimes reserved for the state or pre-emptive rights of the state apply thereby restricting potential foreign investment. Special authorisations are required in some countries for granting mining permits to foreigners. Also, many countries subject foreign direct investment (FDI) in general to review or authorisation processes to pursue diverse policy objectives, including the monitoring of FDI (Norway), increasing the participation of national firms in investment projects (in the recent past in Australia), and by providing leverage measures in cases where national security is deemed threatened. However, recent reforms in some countries (Canada, Australia and Sweden) resulted in lifting some of these restrictions in recognition that economic efficiency was improved by implementing a more liberal FDI regime. As well, within the EC, as internal barriers among Member States are gradually dismantled and policy harmonisation is implemented, fewer restrictions will apply on investments originating from other EC country members.

Environment

It is commonly accepted that mining has potentially harmful and disruptive environmental consequences during prospecting (drilling, sampling), mining (removing over-burden, dust, storage of tailings) and primary and secondary metalworking (emissions, discharges). The sector's activities affect the land (contamination by heavy metals), air quality (emission of particulates and toxic gases, including sulphur dioxide and lead) and water quality (seeping of contaminated water from tailings dams). In order to mitigate the impact of mining activities on the environment and the effects on health and occupational safety, governments have been setting up increasingly sophisticated regulatory systems.

If environmental problems are similar, approaches adopted by governments to address them are very different. The United States environmental policy may be referred to as a "command and control" strategy. The main federal laws lay down objectives to be met in the environmental field. Federal legislation is backed by a set of measures specifying the means whereby the objectives are to be met, implementation being the responsibility of the competent federal agencies. Control, e.g. of water and air quality, is exercised in Canada through a system of federal and provincial permits which involves more informal procedures than those in the United States. In Japan systematic control of the environmental impact of industry began in 1967 with the Basic Law for Environment Pollution Control, backed in 1970 by the Air Pollution Control Law and the creation of the Environmental Agency, the supervisory body in environmental control.

Responsibility for environmental questions often falls with local or decentralised authorities. In Australia and Canada, this means the states or provinces. Through the Canadian Environmental Protection Act, the Canadian Government may monitor pollution, conduct research activities and set targets for atmospheric emissions, but these targets become statutory only if approved by the governments of the provinces. In Australia, the commonwealth government has in recent years legislated on environmental impact assessments, nature conservation, heritage protection and the marine environment, legislation whose purposes have included the protection of environmentally sensitive areas from mining. Work is now underway to harmonise standards between states to eliminate competition in attracting investment. The federal government is seeking a bigger commitment concerning environmental matters in Commonwealth-state Ministerial Councils, which facilitate co-ordination between the two levels of government.

EC countries have generally, until recently, not had environmental legislation specifically relating to the mining and metals industries, apart from general standards, e.g. concerning air emissions (SO_2) and occupational safety in places of work. EC environmental policy has only recently been evolving. The Single European Act, which entered into force in 1987, extended Community competence for certain aspects of the environment.

The main piece of EC legislation to date directly affecting this sector is Directive 85/337/EEC. It stipulates that Member States should undertake environmental impact assessments (EIA) on water, air, the fauna, flora and landscape and ensure the participation of local communities in final decisions on the development of large projects, including mining and metal processing projects. In transposing Council Directive 85/337/EEC, Ireland has adopted statutory procedures requiring a full audit of the environmental impact of projects. Approval of the project must receive the agreement of local communities. In the event of a conflict, the latter are empowered to challenge the decision of the Planning Board through a High Court Action and even through the Supreme and European Courts. In Germany, the government has incorporated the general EC provisions on EIA into national law by means of specific regulations for mining projects.

In the United States, the environmental impact study has de facto developed into a policy measure with broad implications. Approval of companies' projects depends on the outcome of the study. These stipulate not only general standards to be met concerning occupational safety and the siting of the facilities on public lands but also maximum pollution thresholds and the need to return the site to its original condition after the facilities have closed down. Companies operating in Canada are required to undertake an environmental impact study whose content is defined through negotiation between the companies and the government agency concerned.

Quality of air has perhaps been the area of environmental legislation which has most dramatically affected the non-ferrous metals smelting and refining activities. Copper smelting plants, as a main source of SO_2, are subject to strict controls on air quality. The situation may be even more strict in the smelting and refining of primary lead owing to emissions of lead particulates into the atmosphere. For aluminium smelters, air and emission standards have increased costs relative to other producers e.g. in eastern European countries, where fluorine and PAH emissions are still high. Water quality is also of particular concern. In the United States, the capacity expansion of a mine was recently delayed owing to the procedures required for obtaining a water quality permit for a tailings dam.

Waste management is another area of environmental policy with strong potential impacts on the mining sector due, inter alia, to large quantities of acidic wastes from mining and milling. In the United States, solid mining and metal wastes have been excluded from the strict legislation (Resource Conservation and Recovery Act) concerning production, transport, processing, storage and dumping of toxic wastes. Special regulations for this type of waste are to be approved shortly. In Belgium, regional authorities — on whom environmental responsibility has been devolved — have, since 1989, substantially increased the charges for the collection and treatment of waste.

Mining is a high-risk activity, especially in underground mines and health and occupational safety regulations are stringent. The main health problems are caused by the inhalation of dust, toxic gases, high noise levels and prolonged exposure to radioactivity levels. The Canadian provinces have their own legislation controlling health and occupational safety. The standards to be met by each firm are often defined by environmental committees, whose members are both employers and employees, and negotiated in collective bargaining agreements.

Research and development (R&D)

Research and development is an important area of public policy intervention, although programmes in the area of mining and non-ferrous metallurgy are generally not priorities today. Governments may intervene by fully funding basic research, partly or fully financing strategic projects, supplying subsidised loans, loan guarantees and taxation deductions, and assisting co-ordination of various partners in projects. The main objectives of R&D programmes in the countries surveyed are the development of:

- more effective and cheaper geological prospecting and mining technologies and methods as well as new metalworking processes;
- environmentally clean processes with greater flexibility as regards supplies (use of complex or depleted ores);
- reliable recycling processes for bulk waste treatment under environmental control;

— improved quality and effectiveness of metal products with a view to slowing down their replacement in traditional uses as well as developing new applications.

Trade policy and marketing

Ores and concentrates for the six main non-ferrous minerals are generally traded on a duty-free basis, except for the United States which applies low specific duties on lead, zinc and copper. Tariffs on processed and semi-finished non-ferrous metals generally show escalation by successive processing stages in all countries. However, the rate of escalation and the peak rates vary significantly from country to country. The majority of non-ferrous tariff rates are bound against increase in the GATT for all countries, except for Australia where few tariff rates are bound. Tariffs are generally expressed in percentage points and specific rates are applied less frequently. One sliding tariff rate (declining specific rates linked to increasing metal prices) is applied by Japan on refined copper and zinc. All countries except Japan participate in one form or another of regional trading arrangement which provide or will provide for reciprocal duty-free access for non-ferrous products among participating countries. Within the Market Access Negotiating Group of the Uruguay Round of multilateral trade negotiations, tariff levels are expected to be reduced further.

A wide range of non-tariff measures are used by governments to limit trade, such as export and import embargoes, export restrictions, voluntary export restraints, export taxes, licensing, quotas, quantitative restrictions, state trading, domestic price adjustment and certification arrangements. The transparency exercise revealed few examples of non-tariff measures actually in place in the non-ferrous area, but some countries maintain authority to apply such. Although these non-tariff measures do not appear to affect trade flows, it can be argued that the mere existence of restrictive legislative authority exerts a negative effect on trade and investment[7].

Centralised purchasing is seldom used in the non-ferrous metals industry. In France, a metals import group (GIRM) has for many years acted as a central purchasing agency, particularly for copper.

CONSUMPTION/SUPPLY

Countries that are important consumers of ores and metals have traditionally had a specific government policy on security of supplies. Choice of policy instrument varies according to perception of vulnerability and tradition of government intervention.

Securing overseas supply

OECD countries have supported the development of mineral projects abroad to different degrees. In the United States, there are few foreign investment support systems owing to powerful private firms and the size and diversity of the national mining industry. They are confined to taxation measures, e.g. depletion allowances, the funding of equipment purchases by EXIMBANK, guarantees for private loans, investment insurance against expropriation risks (OPIC) and diplomatic support.

Relative to the other countries surveyed, Japan has put in place a more systematic and elaborate supply security policy. Government policy focuses on the promotion of foreign projects aimed at exports to Japan (develop-for-import). Due to the government bearing some of the commercial risks of new mining projects, Japanese firms have been able to meet capital requirements chiefly through bank loans, thereby reducing the share of equity.

Each large consuming country in Europe has its own intervention mechanism for organising supplies, including bilateral aid and preferential treatment of former colonies. Although in the United Kingdom security of supplies is chiefly based on the production and marketing networks of UK mining houses, the government funds several exploration and technical co-operation programmes in Commonwealth countries. The Bureau de recherches géologiques et minières (BRGM) of France intervenes as a technical operator in various programmes of co-operation, and, in certain cases, it participates with its mining branch, COFRAMINES, in the development of ore deposits. In Germany, the main policy instruments are legal and policy guarantees, in particular through bilateral agreements, the promotion of barter and partnerships between German firms and those of developing countries, investment financing, especially for small and medium-sized enterprises and international technical co-operation with the geological services of countries with high mining potential. In Italy, the supply policy is based on exploration work abroad and participation in a few operational projects by state companies. The Spanish Government conducts exploration programmes under bilateral co-operation agreements and R&D co-operation agreements, chiefly in Europe and Latin America. In the early 1980s the EC set up the SYSMIN programme under the Lomé Convention, which aims at helping ACP countries to restructure and modernise their mining sector for certain specific minerals.

Stockpiles

Strategic stockpiling has been pursued by some large consuming countries which feared disruptions in the supply of certain metals. Often these worries concerned metals produced in the former Soviet Union and the southern horn of Africa. The best known is that of the United States. Originally set up in 1939, it consists of 62 materials or groups of materials valued at US$9.0 billion (January 1991 prices). Legislation explicitly states that the stockpile is not to be used for economic or budgetary purposes. Japan has two stockpile programmes, both involving the participation of the private sector; a rare metals stockpile for economic security, and a stabilisation stockpile composed of non-ferrous metals. France began building up a "precautionary" stockpile in 1975 aimed at securing a two month supply of certain materials in the event of supply disruptions. In 1990, the government reaffirmed its strategic interest in maintaining its stockpile.

In view of the significant changes in the geopolitical situation, particularly in the two main producing areas originally targeted, countries are re-thinking their stockpile objectives. The US Department of Defense has recently submitted a report to Congress which sets forth revised stockpile goals of US$3.3 billion to reflect changing national security materials requirements. The United Kingdom and Sweden no longer maintain strategic stockpiles.

Recycling

In recent years recycling has become a priority goal in OECD countries. The development of recycling activities is far from homogeneous in these countries. This is apparent when identifying the firms which have been active in selecting recycling as a new specialisation field and are taking the lead in the organisation of waste flows and the development of waste treatment technologies. As a result, recycling regulations are also at different stages of development.

Several environmental policy measures also impact on this activity, such as dumping charges, waste handling and transport regulations, subsidies for the recovery of materials from wastes, etc. Thus, in certain cases tensions have arisen between the fostering of this source of secondary raw materials and waste management regulations. Industry has been expressing concerns about effects of such regulations on the future development of recycling, which in spite of major technological advances, might be threatened by environmental constraints.

NOTES AND REFERENCES

1. In the latter two countries, however, an exception to the basic principle is that gold and silver mineral rights are vested in the state.
2. In the United States, the Bonneville Power Authority (BPA), also offered the aluminium industry discounts on normal rates on four different occasions between 1983 and 1985 for periods of six months. The system of tariffs indexed to the ingot price for primary aluminium, set up in 1986, expires in 1996 and is not expected to be renewed.
3. The Hydro-Québec supply contracts with thirteen energy-intensive projects were negotiated after 1987 under the "Program of Risk and Profit Sharing". This programme is in line with the industrial policy option for power-intensive development of Quebec. Hydro-Québec has undertaken to deliver 2 100 Mw to these thirteen projects between 1991 and 1998, which according to the Program of Risk and Profit Sharing contracts, would in principle pay the average cost for all Hydro-Québec power plants. However, under the terms of the energy contract concluded between Norsk Hydro and Hydro-Québec, which has been made public, the energy producer is reported to have given rather large discounts. (*Metals Week*, 11 February 1991 and 27 May 1991; *Le Devoir*, 8 May 1991).
4. In Australia, the Victoria state Government negotiated a flexible tariff with the Alcoa for the Portland project, a joint venture involving Alcoa Australia (45 per cent), the Victorian Government (35 per cent), First National Resource Trust (10 per cent) and China International Trust and Investment Corporation (10 per cent). However, because the tariff was calculated using a "normal price" for primary aluminium that was too high (US$1 700/tonne at 1982 prices), metal prices have generally remained below this level and the tariff does not meet normal costs for the power generation supply service.
5. In Norway the situation is quite different, but just as complex. About 85 per cent of the energy produced comes from companies controlled by the local and central authorities. The state-owned power producer Statkraft represented about 30 per cent of the Norwegian production capacity in 1992. One of the most striking features of the electricity market has been the differential in tariffs paid by final users depending on the energy producer and the type of industrial activity. The situation is further complicated by the fact that firms manufacturing energy-intensive products may generate their own electricity from hydro-power sources. For instance, Norsk Hydro with major aluminium interests has now reached self-sufficiency for about 60 per cent of its requirements. In 1991, a new Energy Act entered into force in

Norway, with the aim of reforming the energy market through deregulation and improved competition in the production segment. Historically, state power contracts with power-intensive industry were based on covering the costs of the required investments in a new hydro-power generating capacity. In 1991, the Norwegian Parliament decided to offer a renewal of the contracts expiring in the 1990s, but which would involve a tariff increase in real terms between 1996 and 2010. At the same time, the Parliament permitted Statkraft to re-negotiate contracts with the industry on the basis of normal business terms. This may include linking of electricity prices to the prices of products sold (e.g. aluminium, ferro-alloys and paper pulp) by the electricty user in question.

6. In an agreement with Électricité de France (EDF), Péchiney acquired a holding in a nuclear power plant in exchange for delivery of electricity needs for 25 years at a preferential rate. In 1989, EDF agreed to become joint constructor of a new aluminium plant in Dunkerque. EC competition authorities required the insertion of a clause on the price of electricity supplied which stipulated an increase of 10 per cent per year of the tariff paid by Péchiney between 1997 and 1999.

7. In Australia, export controls are in place for bauxite and alumina, i.e. export prices are reviewed to ensure that fair market prices are received. In Canada, legislation is in place at the federal level and in some provinces to control exports but such rules have never been applied to non-ferrous metals. In Italy, the state has the right to attach special conditions to mineral concessions including conditions relating to the end use of the minerals extracted.

AUSTRALIA

INTRODUCTION

Australia is well endowed with reserves of a wide range of ferrous and non-ferrous mineral commodities including a number of large scale high quality deposits. While the country has been an important producer of base metals and gold since the last century, the Australian minerals industry has experienced a period of particularly significant and sustained growth over the past 25 years. The growth of the Asian market, in particular the steel and base metals industries in Japan and Korea, combined with generally buoyant market conditions has placed the country among the world's main non-ferrous mineral producers. In 1990, the Australian share of world non-ferrous metals production was 37.1 per cent for bauxite, 34 per cent for alumina, 12.0 per cent for gold, 16.5 per cent for lead, 30.0 per cent for ilmenite, 57 per cent for rutile, 53.0 per cent for zircon and 13.0 per cent for zinc. Australia is also a producer of copper and aluminium.

Over the past two decades, the Australian mining sector, including the non-ferrous sector, has increased its contribution to economic growth in Australia. The non-fuel minerals sector contributes around 9 per cent to GDP. Resource based firms dominate the list of top 10 companies in Australia and have proven to be highly competitive by world standards.

The non-ferrous metals industry comprises some 12 main commodities, all of which earned in excess of A$ 150 million in export revenue in 1990-91. The industry is export-oriented and concentrated on a few main commodities: alumina (A$ 2.8 billion), aluminium (A$ 2.0 billion), gold (A$ 3.7 billion) and zinc (A$ 1.1 billion). These minerals account for around 22 per cent of Australia's total exports and some 4 per cent of GDP. Since 1960, Japan has been the largest single destination for Australia's mineral resource exports. The importance of Japan has declined slightly in recent years reflecting growing trade with newly industrialised Asian countries.

Recent Australian policies have been directed at maximising the return from the mineral sector by facilitating economic and environmentally responsible development. The goal of recent government policy initiatives has been to improve the international competitiveness of the minerals sector, improving market access through multilateral and bilateral trade negotiations and ensuring sustainable development while providing an appropriate return to the community for the use of its mineral resources.

Few restrictions are imposed by government on the mining industry. From February 1992, overseas investment in new mining projects is no longer restricted to 50 per cent ownership. Foreign investment guidelines have been administered flexibly in recent years. Export controls also apply to a few selected commodities.

Australia's minerals sector expanded rapidly in the late 1970s and early 1980s. The decline in mineral and metal prices during the 1980s resulted in a rationalisation of the minerals sector. Productivity increased and debt levels were reduced leaving the sector in a much stronger competitive position. Minerals make the largest contribution to exports and is a key sector in overcoming Australia's current account and external debt problems.

The government's policy response to foreign trade deficit and overseas debt problems has essential macroeconomic and micro-economic elements. On the macro-economic level, the government has deregulated the Australian financial market, while the foreign investment policy has been progressively liberalised to facilitate overseas investment. On the other hand, tax reforms were implemented to provide an environment conducive to both productive investment in Australia and direct investment overseas.

These measures were complemented by the development of a strategic approach to industry, with science and technology policies that operate on two fronts: first, this strategy seeks to stimulate improvements in industry performance and to provide the best environment in which internationally competitive firms and industries can grow; second, it seeks to integrate Australian firms with the world economy and encourage them to search for international opportunities.

Recognition of the economic importance of the minerals sector has been reflected in recent government reforms aimed at reducing the cost of inputs used by export industries. General reductions in tariffs (generally phasing to 5 per cent by 1996) and transport reform, for example, will reduce the cost of imported equipment and improve operating efficiency.

Organisation of relevant government agencies

Australia has a federal system of government comprising six states, a self-governing Territory and a federal Commonwealth government. Both state and Commonwealth governments have an important role in the development and marketing of minerals and metals.

Under the Australian Constitution, legislative powers were not specifically assigned to the Commonwealth and therefore primary responsibility and control over mining rests with state and territory governments. The Commonwealth government's responsibilities cover general economic management, taxation, international relations, foreign capital and investment, but also extend to regulation of exports and aboriginal affairs.

Each state and the Northern Territory has mining legislation and administrative arrangements governing the granting of mineral exploration and mining titles. State governments, by virtue of their wide powers to regulate matters within their own boundaries, are directly responsible for the day-to-day administration, including exploration and development titles, supervision of mining and associated operations, compliance with health and safety requirements and environmental protection, and the levying of royalties and other like charges.

In several states, major reviews of mining laws have been conducted since 1988 in an attempt to simplify and streamline the regulation of resources development and to integrate mining laws more effectively with other legislative schemes which affect mining, such as planning and environmental protection laws.

There are two federal government ministers responsible for the primary industries and energy portfolio. The Minister for Primary Industries and Energy is responsible for the overall operation of the portfolio and the development and execution of portfolio policy. The Minister for Resources takes responsibility for a number of specific issues and policies within the portfolio.

The Department of Primary Industries and Energy (DPIE) is the principal Commonwealth portfolio with responsibility for policies affecting the mining industry. The Department provides policy advice, scientific as well as economic analysis and administration of programmes related to primary industries including mining. The objective of the Department is to promote an internationally competitive minerals industry consistent with ecologically sustainable development principles and one which provides an appropriate return to the community from the use of community resources.

DPIE has two bureaux dealing specifically with mining, the Bureau of Mineral Resources (BMR), the geoscientific research and resource assessment organisation, and the Australian Bureau of Agricultural and Resource Economics (ABARE). The role of ABARE is to enhance the economic performance of the Australian primary and energy industries by providing independent and objective economic analysis and assisting in government policy development and decision-making.

Co-ordination of Commonwealth and state policies is provided for through the Australian and New Zealand Minerals and Energy Council (ANZMEC). Its objectives, endorsed by the Premiers' Conference in 1976, are (i) to facilitate economically competitive development of Australia's minerals and energy industries; (ii) to promote environmentally responsible mineral and energy industry development; (iii) to improve the co-ordination and where appropriate the consistency of policy regimes across Australia; (iv) to encourage new and expanded investment in potentially competitive minerals and energy development opportunities; (v) to provide an opportunity for information exchange and a forum for policy discussion.

The Industry Commission (formerly the Industries Assistance Commission — IAC), has a statutory responsibility to report publicly on the assistance provided to industries by the government and the effects of that assistance on the development of industries and the economy as a whole. The Commission's measurement of industry assistance has largely concentrated on forms of assistance affecting trade, such as tariffs, quotas and production bounties.

The forms of assistance usually analysed are three-fold:

i) state assistance arrangements: price controls, royalty concessions for local use;

ii) Commonwealth taxation measures: tariffs on materials, tariffs on capital, capital depreciation arrangements, excises on fuel, export duties;

iii) other measures, such as regulations, and export restrictions.

A pilot study by the IAC from 1988[1] found that the mining and mineral processing industries in Australia generally receive low or negative effective rates of assistance when compared to agriculture and other manufacturing sectors.

The Industry Commission (IC) recently held a public inquiry into mining and minerals processing in Australia. In its final report released in April of 1991, a wide range of issues of importance to the industry were addressed and recommendations for government action made. Major issues identified related to lands access, environmental concerns, taxation and royalties, construction and operating costs.

Effective rates of assistance reported by the IC for non-ferrous metals in 1988-89 were -1.9 per cent for bauxite, -3.9 per cent for copper ores, -2.7 per cent for gold ores, -2.1 per cent for nickel ores and -2.9 per cent for silver/lead/zinc ores. The negative rates of assistance generally indicate that the initial impact of border assistance is to raise mining costs above international levels such as through tariffs on inputs.

Legislation to establish the Resource Assessment Commission (RAC) was introduced in 1989. The RAC is designed to provide a vehicle for inquiring into, and reporting to the Government on, the environmental, cultural, social, industrial, economic and other aspects of Australian resources and their uses.

Private industry groups

Despite the overall healthy state of the Australian minerals and metals industry, the private sector often expresses the concern that government controls sometimes discriminate against the industry in areas like taxation, foreign investment, land use and the environment. As a result, a number of groups have been formed in order to defend the interests of the minerals and metals sector. Among these groups, the Australian Mining Industry Council (AMIC) is perhaps the most active in influencing non-fuel mineral policy.

The big Australian companies in the 1980s broadened their interests across a wider geographic area and spread of metals and minerals-related activities to reduce their exposure to commodity price volatility and to Australia's economic conditions. Australian companies with substantial international mining interests include: BHP (Broken Hill Proprietary); CRA; MIM (Mount Isa Mines); and North Broken Hill Peko Limited. This international strategy has taken them into the major market areas of the world — East Asia, North America and Europe — as participants in the metals and minerals business, from mining to the actual consumer of metal products.

PROSPECTING AND EXPLORATION

Geological services

Exploration for non-ferrous metals is primarily a private enterprise activity in Australia. However, governments play a supporting role by conducting research and providing the maps and other basic information which help exploration companies define prospective regions.

The Bureau of Mineral Resources (BMR) has the role of developing an integrated, comprehensive, scientific understanding of the geology of the Australian continent, the Australian offshore area and the Australian Antarctic Territory as a basis for mineral exploration. The BMR is the primary national source of geoscience data. In recognition of BMR's new emphasis on environmental geoscience, one of BMR's functions is to help establish and co-ordinate the national geoscience knowledge base required for the consideration of resources use, land use and environmental issues.

In 1989, a National Geoscience Mapping Accord was established to promote the co-operative use of state, territory and Commonwealth resources in the mapping needed to provide a basis for exploration and assist in decision making on resource and land use issues.

Lands availability

In Australia, access to land by the mining and minerals processing industries has traditionally been controlled by state governments but, more recently, the Commonwealth government has also passed legislation which impacts on the mining industry's access to land. As a result the mining industry is of the view that access to land in Australia is not only unduly restricted but also subject to an unnecessarily complex system of land use regimes. At the same time, conservation groups are of the view that more land should be set aside from mining in the form of nature reserves and national parks.

In Australia mineral resources are Crown Property and they are administered by the state and territory governments on behalf of the people. The states and the territory therefore generally have the right to grant exploration and mining rights and lay down royalties and other duties.

The Commonwealth and several states have recognised that aboriginals should have the right to control what happens on their land. While the states are responsible for aboriginal legislation in their area, the federal government is responsible for it in the Northern Territory. Under this legislation aboriginals are able to control access for the purpose of exploration as well as to negotiate compensation directly for damage resulting from exploration and, in the Northern Territory, payment from mining activities. This does not, however, apply in all jurisdictions. Further, both federal and states/territory legislation allows the identification and protection of sites of significance to aboriginal heritage (whether on aboriginal land or not). Aboriginal areas represents 11 per cent of the total land in Australia.

Arrangements for allowing access to Commonwealth land for exploration and mining are enacted through the Lands Acquisition Act 1989. The Act provides for regulations to be developed to improve the process through which access can be gained. The principle of multiple land use, embodied in the 1984 National Conservation Strategy of Australia (NCSA), where such use is not inconsistent with the primary use of the land concerned (rural areas, pastoral leases), is a central feature of the government's broad decision-making principles. These principles translated into the following guidelines and outlined in the Prime Minister's 1989 Statement on the Environment:

- there should be an integrated approach to conservation and development by taking both conservation and development aspects into account at an early stage;
- resource use decisions should seek to optimise the net benefits to the community from the nation's resources, having due regard to efficiency of resource use, environmental considerations and an equitable distribution of the return on resources; and
- Commonwealth decision, policy and management regimes may provide for additional uses that are compatible with the primary purpose values of the area, recognising that in some cases both conservation and development interests can be accommodated concurrently or sequentially, and, in other cases, choices must be made between alternative uses or combinations of uses.

Authority for exploration licenses and mining leases

The ownership of onshore and coastal mineral rights rests largely with state and territory governments. The Commonwealth has ownership in offshore areas and of uranium in the Northern Territory.

The various state governments have the responsibility for the granting of exploration rights and mining leases, the approval of mining operations, the regulation of health and safety legislation, and the levying of royalties and other like charges. Administrative arrangements covering the granting of minerals exploration vary somewhat from state to state.

In most states, exploration licenses are issued on a non-price, "first come, first served" basis and carry a range of conditions which may include the amount of expenditure to be undertaken and a minimum annual work programme. Mining leases are usually of limited tenure and involve nominal lease payments to the relevant government often on a per hectare basis.

In all states, the holder of an exploration licence has priority in obtaining a mining lease over the area covered by that licence, provided that the proposed mining operation is acceptable to the state government (e.g. according to environmental considerations and other legal dispositions). Two types of mining leases are usually available: those issued generally under the various Mining Acts, and specific project agreements which are negotiated directly with the companies involved. Leases granted under specific project agreements may require the company to provide infrastructure or to undertake further processing within the state in which the mine is located.

Taxation incentives

Exploration is primarily a private enterprise activity with governments mainly playing a supporting role. Mining and exploration companies are subject to federal income tax. Mining exploration expenditure is deductible from income derived from any source in the year incurred, and losses incurred by a resident company are transferable to another resident company in the same group, provided there is 100 per cent common ownership. In cases where there is either no assessable income or a company is not part of a group then no immediate assistance is obtained, but expenditure may then be carried forward indefinitely and offset against future income.

MINING AND METALS PRODUCTION

Mining

Infrastructure support

Mineral developments require a significant level of supporting infrastructure. In some cases, existing infrastructure can be adapted and used to service mining developments but many recent large scale mining and minerals processing projects have required the establishment of large and dedicated facilities.

Infrastructure services such as electricity and water, port and loading facilities, communications, road and rail transport, health and education services are the responsibility of State governments. Policies on provision of social and industrial infrastructure vary considerably amongst states. While the private sector has funded most industrial and community/social infrastructure in remote areas, state governments and their agencies are major providers of services such as rail, ports and electricity.

Environment

Mining and minerals processing in Australia face a number of regulations applying to their operations. State/Territory governments have prime responsibility for environmental protection and rehabilitation of mining projects. Regulations are enacted either through each State/Territory Mining Act, determining the terms and conditions attached to the issue of mining leases, or in most cases, under separate environmental legislation. Local government responsibilities focus on detailed aspects of land planning and environmental protection (including for example, tree preservation and sewerage).

The Commonwealth government is able to considerably influence environmental protection when it has a specific role in the decision making process. A summary of the respective Commonwealth/state responsibilities affecting mining is shown below.

Table I. **Australia: Government interventions affecting mining activities**

Interventions	Bauxite	Tin	Copper	Zinc/Lead	Silver	Gold
Royalties	SC	SC	SC	SC	SC	SC
Federal Income Tax	CC	CC	CC	CC	CC	CC
Export Controls	CC					
Controls on Foreign Investment	CC	CC	CC	CC	CC	CC
Coastal Shipping	CC					
Excess rail freights				SC	SC	
Aboriginal land rights	SC&CC					

SC = State control;CC = Commonwealth control
Source: Industries Assistance Commission (1988), updated by DPIE.

The Commonwealth export control powers for non-ferrous minerals and metals only apply to bauxite/alumina. Exporters are required to seek export approval on the completion of negotiations to ensure that a fair market price is received. This was the case, for example, of the Alusuisse-Lonza Group that has obtained from the government in 1990 the authorisation to export 40 million metric tonnes of bauxite over the next 20 years. These controls do not, however, involve the payment of any export duty.

Minerals taxation

The minerals industry in Australia makes a significant contribution to the revenues of governments. Besides the Commonwealth income tax, the industry is subject to state royalties and other resource related charges (e.g. infrastructure charges) that are not uniform from state to state. The 1988 economic statement announced the reduction of the corporate tax rate from 46 per cent to 39 per cent. In 1988-89 the sector paid over A$ 4.6 billion in royalties, charges and taxes.

There are currently three basic types of mining royalties schemes used in Australia: specific, *ad valorem* and profit based. Specific and *ad valorem* royalties are the most commonly used and apply to most minerals across all states except the Northern Territory. Specific royalties are levied on a per unit of production basis usually involving a flat rate per tonne of ore, concentrate or contained metal. *Ad valorem* royalties are levied as a percentage of gross value. A proportional profit royalty, levied on a proportion of accounting profit, is applied to most mining operations in the Northern Territory.

Table II. **Australia: Government royalty schemes applied to the mining industry**

Royalty scheme	Royalty base	Applicable minerals
New South Wales:		
Specific royalties	Volume of output	Bauxite and alumina
Ad valorem royalties	Value of output	Most other non-ferrous minerals
Profit royalties	Accounting profit	Lead, silver, zinc and copper mined at Broken Hill and Cobar
Queensland:		
Specific royalties	Volume of output	Nickel, silver, lead, zinc, and copper mined at Mount Isa
Specific/*ad valorem* combined	Volume/value of output	All other non-ferrous minerals
South Australia:		
Ad valorem royalties	Volume of output	All minerals
Tasmania:		
Ad valorem Profit royalties	Value of output/ Accounting profit	All minerals
Victoria:		
Ad valorem royalties	Value of output	All non-ferrous
Western Australia:		
Ad valorem royalties	Value of output	Most non-ferrous minerals
Northern Territory:		
Proportional profit royalties	Accounting profit	All minerals (except uranium)

Source: Industries Assistance Commission (1988), updated by DPIE.

All states have a royalty system and elements of the three schemes are used in combination in some instances:

i) New South Wales collects its royalties under the authority of the Department of Mineral Resources, based either on a specific or *ad valorem* basis, on quantity or value of mining production. Specific royalties of A$ 0.35 per tonne apply to bauxite and A$ 0.70 per tonne to alumina. Most other non-ferrous minerals are treated on an *ad valorem* basis and rates do not exceed 4 per cent of the minerals value. An exemption to this system exists in the cases of the silver-lead-zinc mines at Broken Hill (20 per cent profit royalty based on average ore grade and mining depth), and the copper mine at Cobar where royalties are levied on a progressive scale as a percentage of the profits received from their sale.

ii) Queensland levies royalties on the basis of quantity, value or profits under the jurisdiction of the Department of Resource Industries. Most of the non-ferrous metals are submitted to a combined scheme of royalties usually the greater of 2 per cent *ad valorem* or 5 per cent profit. The exceptions to this scheme are bauxite which is the greater of 10 per cent *ad valorem* or a specific rate, and silver, lead, zinc and copper mined at Mount Isa which is at a specific rate of no less than A$ 1.00 per tonne.

iii) South Australia has simplified its royalty collection with an across-the-board tax of 2.5 per cent of the current mineral value upon its extraction. The exception is a combination 2.5 per cent *ad valorem*/profit royalty for the gold, copper, uranium mine at Olympic Dam. The Department of Mines and Energy is the authority responsible for royalty collection.

iv) Tasmania's Department of Resources and Energy collects royalties on a proceeds/profits basis. A combination 1 per cent *ad valorem*/profit royalty applies to all non-ferrous metals produced except for the Hellyer (silver, lead, zinc) mine which is 2.5 to 3.5 per cent *ad valorem* plus a profit component.

v) The Victorian Department of Manufacturing and Industry Development is responsible for collecting royalties. Victoria currently uses a 2.75 per cent *ad valorem* royalty for all non-ferrous metals except for gold which is exempt.

vi) Western Australia generally collects royalties on an *ad valorem* basis, except for gold which is exempt. Rates usually vary between 2.5 per cent of realised value for tin concentrate to 7.5 per cent for bauxite. The net effect is to approximate a 10 per cent *ad valorem* on the ex-mine ore value. A special arrangement exists for the Argyle (diamond) mine which uses the greater of 7.5 per cent *ad valorem* or 22.5 per cent profit.

vii) The Northern Territory Government uses an 18 per cent accounting profit royalty for profits over A$ 50 000. This applies to all mineral projects that commenced after 1982 and for those that elected to change from the previous royalty arrangements (generally 1.25 per cent of ex-mine production value).

Gold mining benefited for many years from a total tax exemption from income taxation. The exemption was removed on 1 January 1991. A company's gold mining income is now treated the same as other income. To ease the impact of the decision, gold mining companies are allowed to carry forward capital expenditure incurred before 1991 and deduct it from their taxable income after that date.

Generally, current expenditures necessary to earn income are tax deductible in full in the year in which the expenditure is incurred. Certain capital expenditures by mining companies are covered by a special provision of the Income Assessment Act which allows qualifying expenditure to be deductible by instalments over the lesser of mine life, or ten years beginning in the year in which the expenditure was incurred.

Categories of expenditure covered under the special mining provision include site preparation, certain buildings including housing, water, light and power facilities, plant for exploration and buildings for certain treatment processes.

Metals production

Inputs

Electricity

Electricity costs are a significant cost component of non-ferrous metal processing, particularly for aluminium smelting. The electricity supply industry in Australia is organised on a state basis. Government authorities have control of generation and supply in each State. Industrial users of electricity take their supplies under contracts entered into with supply authorities. Such contracts are designed to reflect specific users' requirements. For example, in return for lower tariffs, a contract may contain clauses which enable supplies to be interrupted at times of peak load or when other supply difficulties occur. Contracts may also reflect the ability of large users of electricity such as aluminium smelters to guarantee a significant demand over the long run.

Australia's availability of significant coal resources as a base to generate competitively-priced power for smelting and refining has been critical to the establishment of new processing ventures. However, although three additional aluminium smelters were commissioned in Australia, the rates of growth of the economy and electricity demand proved to be lower than forecast. This factor contributed to the provision of a much higher level of generating capacity than required and to an increase in costs.

In some states, there are proposals for the private sector to have a more significant role in the electricity supply industry. In 1988, for example, the State Electricity Commission of Victoria (SECV) announced an 'incentive' scheme to attract up to 150 megawatts (Mw) of additional co-generation capacity. CRA has put forward proposals to electricity authorities for the establishment of a generation project in New South Wales. In Western Australia it has been decided that the next increment of generation capacity will be provided by a 600 Mw privately owned coal-fired power station. In Queensland, the government is negotiating the sale of the Gladstone power station to the aluminium producer Comalco, a subsidiary of CRA.

In May 1991, the Industry Commission reported on its inquiry into Energy Generation and Distribution, recommending major structural changes so as to increase competition and efficiency in the electricity supply industry. A National Grid Management Council is to be established to co-ordinate the sound economic and environmental development of the electricity supply industry in South East Australia. The Electricity Commission of NSW has announced a major restructuring of its activities to increase efficiency, the restructuring will see its generation activities grouped into three competing business units.

Rail freight charges

Some state governments such as those of New South Wales and Queensland are able to apply discriminatory pricing policies on the provision of state-owned transport services because of their legislated monopoly powers. Public rail authorities collected over A$ 1.3 billion in freight revenue in 1986-87, that is to say, over 60 per cent of total freight revenue collected by public rail in the period.

In the non-ferrous area, excess rail freights were considered by the Industries Assistance Commission[2] as affecting mining activities of zinc, lead and silver, thereby acting economically as a tax on such activities.

Waterfront, port services and shipping costs

The minerals sector is the single major user of waterfront and port services, as minerals represent around 60 per cent of total coastal shipping cargoes and bulk users (mainly minerals) have been increasing their share of total expenditure on waterfront services from 40 per cent in 1980-81 to 53 per cent in 1986-87.

The Commonwealth government recognised the waterfront problem and referred the matter to the Inter-State Commission in 1986. Problems identified were: low productivity, high cost structures, inflexible work practices and conditions, lack of commercial orientation by port authorities, costs imposed without being based on services provided, inefficient management, unreliability and excessive delays.

The Commonwealth's three year waterfront reform programme, commenced under the guidance of the Waterfront Industry Reform Authority (WIRA) in 1989. The programme aims at reducing the stevedoring workforce by at least a third (target of a total 3 000 to go by September 1992). It will also produce significant improvements in productivity and efficiency through fundamental changes in employment arrangements and work practices. Under the In-Principle Agreement (IPA) the government has committed A$ 154 million on a dollar for dollar basis with industry, for early retirement/redundancy packages, training and workplace restructuring.

The cost of shipping commodities around the Australian coast is relatively high by world standards. Under the Navigation Act of 1912 coastal shipping is restricted to licensed vessels which comply with Australian standards of manning, accommodation and wage levels. This effectively eliminates competition from overseas vessels and limits shipping services to higher cost Australian vessels. Importing foreign ships is also restricted.

These regulations operate to increase the price of coastal shipping services, specially in the case of bauxite and alumina. Thus, for example, to move its alumina from Gladstone to its smelter in Tasmania, Comalco time charters an Australian vessel. Estimates of the savings if the vessel operated under conditions competitive with foreign vessels approximate A$ 5 million per annum[3].

The Commonwealth initiated a three year programme of coastal reform in June 1989. The programme includes reductions in crew sizes. To the extent that several minerals companies, e.g. BHP and Pasminco, own and operate their own vessels, the current reforms have led to immediate cost savings.

Input tariffs

Tariff protection affects many of the inputs used by mining companies by increasing the price of imported goods competing on the local market. In 1988, the Commonwealth government decided to implement a programme of tariff phase-down to a new two-tier general level for tariffs of 10 per cent and 15 per cent which was implemented on 1 July 1992. The government also granted duty-free entry to mining and minerals processing equipment not made in Australia. Further tariff reductions were announced as part of a package of initiatives to improve industry competitiveness in March 1991. General tariffs at the rate of 15 or 10 per cent by July 1992 will be phased down to a general tariff level of 5 per cent by July 1996. Also of benefit to the non-ferrous metals sector was the decision to broaden the wholesale sales tax exemptions to cover a wider range of inputs to mining such as mining railways.

Direct and indirect support

Financial aid is a policy measure open for use by the States, and to a lesser degree, by the Commonwealth government. The states, for the most part, are responsible for the various programmes designed to promote regional development that award grants, loans and subsidies to the mining industry, although the Commonwealth also contributes funds to the states to be utilised in their respective programmes.

State governments in Australia have used various forms of assistance to facilitate the development of mining operations. The viability of Western Mining's Kambalda nickel operations, for example, was ensured by the Western Australia government's grant of exploration titles for a certain period. The government also dedicated a specific quantity of water and provided the land for the township.

State governments are able to negotiate directly with mining companies to determine project specific royalty regimes, usually for the large mining and minerals processing projects. In return for the Kwinana refinery, the Western Australia government agreed to fix royalties at 2 per cent of the value of contained nickel for five years. A 10 per cent discount on rail freight and fuel oil was also granted in return for the mining company's purchase and maintenance of rolling stock to transport nickel products[4]. Land for the Kwinana Nickel refinery was obtained by the Western Australian government's reclaiming private land and selling it to Western Mining, with additional Crown land, for the cost of reclaimed land.

HORIZONTAL POLICIES

Foreign investment controls

The Australian Government welcomes foreign investment in the non-fuel minerals industry subject to certain conditions. The Commonwealth's foreign investment policy has the broad objective of encouraging forcign investment consistent with the needs of the Australian community. The Foreign Investment Review Board (FIRB) was established in 1976. The FIRB is an advisory body. The chairman and members have a wide range of experience and provide the government with a business-oriented and independent source of advice on foreign investment matters.

The government recently removed the requirement of 50 per cent Australian ownership of mining projects (excluding uranium). Proposals for new investment in mining of A$ 10 million or more still need to be ratified by the FIRB. Mining projects may proceed unless they are regarded as contrary to national interests.

Australian authorities from time to time have exercised their power to control foreign investment in a manner meant to accomplish other objectives, as well. For example, the Commonwealth in 1987 used its foreign investment powers to reject a proposal to mine silica at Shellburne Bay in Queensland for environmental reasons instead of foreign investment considerations.

In 1988, the New South Wales Liberal Government announced the removal of the previous requirement of 51 per cent of Australian equity under the Foreign Investment Guidelines for mining ventures.

Environment

Environmental planning and management has become an increasingly important aspect of the arrangements for management of exploration and mining tenements. Development legislation and the associated licensing systems have historically incorporated certain provisions for environmental protection and have been used increasingly for this purpose in recent years. Nevertheless, the main measures for environmental protection are embedded in a separate body of specific environmental legislation, policies and administrative procedures. These measures are sometimes used in conjunction with or implemented through the licensing system established within development legislation.

This body of environmental protection measures comprises land use planning; environmental impact assessment (EIA) of new projects and major expansions of existing projects; legislation protecting air, land, water quality and controlling pollution, waste disposal and handling of hazardous substances; and reservation of land for specific environmental purposes, chiefly the protection of recreation and nature, and conduct of specific research.

Responsibility for environmental protection is generally a state/territory responsibility, although the Commonwealth is able to exercise influence or direct powers when it has a specific role in the decision-making processes (such as in the case of foreign investment proposals or projects and activities subject to export controls) or when international responsibilities are involved (such as activities in World Heritage areas).

Complementary environmental assessment procedures have been agreed between the Commonwealth and the states to avoid duplication of actions. The Commonwealth and the states are currently looking at reinforcing co-ordination of project approvals processes. Accordingly, most aspects of mining activities are subject to environmental controls in one way or another.

Projects are generally subjected to monitoring by the relevant State Environmental Protection Authority or officers of the Mines Department or both in relation to specific environmental conditions of approval. Rehabilitation requirements are normally specified under the terms and conditions of the approval and all States now require rehabilitation performance guarantees for new projects.

Research and development (R&D)

The Australian Government has implemented a broad set of policies and programmes designed to encourage business expenditure on research and development, and which benefit the non-ferrous mining sector. First, funds are directly allocated to the Commonwealth Scientific and Industrial Research Organisation (CSIRO). The Organisation conducts scientific and technological research relating to the more effective definition, utilisation and management of Australia's earth resources. The work of CSIRO is closely aligned with the strategic needs of industry with a strong emphasis on delivery of benefit in the market place with pay-offs to companies and to the community. There is now a requirement that CSIRO achieve 30 per cent of its funds from external sources, thus providing greater incentive to enter into collaborative research ventures with mining companies on a commercial basis. CSIRO has contributed successfully to the development of a new smelting process for non-ferrous metals which is now being marketed commercially overseas.

Second, to assist in the continuing task of updating technology available to the industry, and in support of those companies unable to take advantage of the tax concession (such as start up companies not yet incurring a tax liability), the Commonwealth government provides grants to companies for industrial research and development projects through the Australian Industrial Research and Development Incentive Board.

More importantly, companies are also eligible for the 150 per cent tax incentive programme for industrial R&D. To encourage companies to attain a meaningful level of R&D, the full 150 per cent deduction is available only if a company's total R&D expenditure for the year exceeds A$ 50 000. For annual expenditures between A$ 20 000 and A$ 50 000, the concession operates on a sliding scale with no concession available for expenditure below A$ 20 000 per annum. Losses arising from R&D expenditure can be carried forward against future income. To encourage companies requiring small scale R&D to take their problems to research centres, these minimum thresholds do not apply to R&D contracted to approved agencies, such as most government research centres.

The March 1991, Government Industry Statement, announced the extension of the R&D tax concession for an indefinite period at a rate of 125 per cent, commencing in 1993, when the existing 150 per cent rate is due to expire.

The mining industry also benefits from government funding through the BMR, CSIRO and ABARE. Nevertheless, roughly two-thirds of total mining R&D is funded by the private sector with the greater proportion of such activity actually being undertaken by public institutions like CSIRO.

The Australian Minerals Industry Research Association (AMIRA) research brokering mechanism plays an important role in fostering generic pre-competitive research. AMIRA was registered under the Companies Act of Victoria in 1959 with the aim of promoting the economic recovery of minerals through the development and adaptation of the relevant science and technology. Membership of AMIRA is open to companies interested in the services it provides. The Association does not undertake research but it initiates and co-ordinates jointly-sponsored R&D contracts on behalf of its members. Research performed by CSIRO accounted for around 28 per cent of AMIRA's A$ 9 million budget in 1990.

Trade policy and marketing

The Australian Constitution grants the Commonwealth government power over international trade. However, these powers are rarely exercised in any direct way. The minerals industry, like other Australian exporters, is eligible for a range of services offered through the Australian Trade Commission (AUSTRADE). These include: taxable cash grants for eligible expenditure on certain export promotional activities through the Export Development Grants Scheme; and a range of insurance, guarantee and finance facilities designed to complement services from the private sector through their export insurance arm, the Export Finance and Insurance Corporation.

In addition, the Commonwealth has the constitutional authority to levy duties on exports. The objective of such duties is to encourage domestic processing; the level of duties would decrease as the degree of processing increases. No such levies exist, however, on any of the non-ferrous minerals/metals exported, and the application of such duties has never been seriously considered as a policy measure by the Commonwealth government.

Since 1979, the number of commodities subject to export controls has been reduced from over 50 to the current 4, which includes only bauxite/alumina and mineral sands from the non-ferrous sector. Controls on bauxite/alumina allow the government to monitor pricing arrangements between vertically integrated companies. Those on mineral sands have been retained for environment and heritage reasons, but, the export control process was simplified to control a total volume of output over a given time frame.

In 1990, import tariffs on ores and unwrought metals were reduced to duty-free entry from 2 per cent. More recently, the Commonwealth government decided to phase down across the board all tariffs as an attempt to stimulate domestic competition and to reduce cost pressures. On 1 July 1992, all tariffs were reduced to either a 10 or 15 per cent level. The rate of duties will be further reduced in annual steps to a uniform 5 per cent by 1 July 1996. This unilateral tariff reduction programme is also considered as an advanced contribution of Australia to the ongoing Uruguay Round of trade liberalisation negotiated under the auspices of the GATT.

In 1983, the Australia-New Zealand Closer Economic Relations Trade Agreement (CER) came into effect which provided for the progressive elimination of barriers to trade between the two countries. The last remaining border restrictions on bilateral trade, covering all products including non-ferrous products, were removed as of July 1990. The bilateral trade in non-ferrous remains relatively small.

CONSUMPTION/SUPPLY

Recycling

The recycling of non-ferrous metal products in Australia is primarily in the hands of the private sector although policies at different levels of government have an impact on recycling levels.

State governments influence recycling through their environmental and industry development initiatives, their own purchasing policies and their controls over local government. The Victorian Environment Protection Authority has implemented "voluntary" recycling targets for certain beverage containers and is considering targets for other products while South Australia has implemented container deposit legislation taking the form of discriminatory deposits on different containers.

The Commonwealth government is finalising a National Waste Minimisation and Recycling Strategy which was expected to be released mid-1992. The strategy focuses on policies aimed at reducing waste produced at source but also includes a range of measures formulated to promote recycling. The major non-ferrous metals covered by the strategy are aluminium and heavy metals including lead and mercury.

Australia has one of the world's highest aluminium recycling rates and the Commonwealth government is committed to maintaining this position. The aluminium industry has nominated an indicative 65 per cent recycling rates for used beverage containers by 1995 compared with the current recycling rate of 62 per cent. Lightweighting has contributed an additional 30 per cent to waste reduction since 1968 and further gains are anticipated from an ongoing R&D programme.

The Commonwealth government recognises the need to minimise the amount of heavy metals entering the environment and is developing, in conjunction with industry, a scheme to promote a 100 per cent recycling rate for lead-acid batteries. The feasibility of recovering heavy metals contained in dry-cell batteries will also be investigated.

NOTES AND REFERENCES

1. Industries Assistance Commission: *Assistance to mining: some issues and preliminary analysis*, Australian Government Publishing Service, Canberra, January 1988.
2. Industries Assistance Commission: Inquiry into Government (non-tax) Charges, Pricing and Production of Government-provided Goods and Services, Information Paper No. 3, Australian Government Publishing Service, Canberra, September 1989.
3. Australian Mining Industry Council: The Mining Review, June 1986.
4. The Nickel Refinery Agreement, Western Australia Government Act No. 24, 1968.

BIBLIOGRAPHY

AMIC (Australian Mining Industry Council) (1988), *The problem of access to land*, April.

AMIC (1989), *What mining means to Australians.*

BARNETT, D. W. (1988), "The Minerals Industry in Australia", *The Minerals Industries of ASEAN and Australia*, McKern, B. & Koomsup, P. (eds.), Allen & Unwin, Sydney.

Bureau of Mineral Resources, Geology and Geophysics (1989), "Australian Mineral Industry", *Annual Review for 1987*, Australian Government Publishing Service, Canberra.

CHARLTON, P. (1987), "Foreign Investment — The Chasm between the Policy and the Act", *The Mining Review*, AMIC, November.

Department of Resources and Energy (1986), *An Overview*, Mineral Resource Development Series.

FREEBAIRN, J. (1987), "Preservation and Conservation Versus Mining Development: Some Economic Issues", *The Mining Review*, AMIC, May.

HAWKE, R. (1989), *Our Country our Future*, Statement on the environment by the Prime Minister of Australia, Australian Government Publishing Service, Canberra, July.

Industries Assistance Commission (1988), *Assistance to mining: some issues and preliminary analysis*, Australian Government Publishing Service, Canberra, January.

Industries Assistance Commission (1989), *The Electricity Supply Industry in Australia, [a study undertaken as part of the inquiry into government (non-tax) charges]*, Australian Government Publishing Service, Canberra, September.

KERIN, J. (Minister for Primary Industries and Energy) & COOK, P. (Minister for Resources) (1988), *Primary Industries and Resources — Policies for Growth*, Australian Government Publishing Service, Canberra, May.

KERIN, J. (Minister for Primary Industries and Energy) & COOK, P. (Minister for Resources) (1989), *Research, Innovation and Competitiveness*, Australian Government Publishing Service, Canberra, May.

KEYES, R. & NOSKOWSKI, A. (1990), *The Role of Mining and Mineral Processing in the Australian Economy*, Paper presented on the UNCTAD/ESCAP Seminar on Mining, Minerals and Economic Development in the Asia/Pacific Region, Perth, Australia.

McINTOSH, L. (1987), "Land Access and the Shrinking World of Mining", *The Mining Review*, AMIC, November.

McKERN, B. & WALTHO, A. (1988), "Australia", *Minerals Processing in the Industrialisation of ASEAN and Australia*, McKern, B. & Koomsup, P. (eds.) Allen & Unwin, Sydney.

TRACE, K. (1987), "Coastal Shipping — An Impediment to Adding Value to Mineral Exports", *The Mining Review*, AMIC, May.

BELGIUM

INTRODUCTION

Belgium no longer has any non-ferrous ore mines in production. A substantial deposit of calamine (a zinc oxide) contributed to a large extent to the development of the non-ferrous metals industry. From the early 19th century, this deposit, located west of Aachen, was the source of zinc production in the Liège Basin and in the Campine area. The zinc also resulted in the production of allied metals such as lead and silver.

As early as 1837, Société Générale de Belgique was closely connected with the sector when it helped to set up Société des Mines et Fonderies de Zinc de la Vieille Montagne, which took over and developed the Liège zinc works.

The key role played by Union Minière is also to be noted. This company was the successor of Compagnie du Congo pour le Commerce et pour l'Industrie, which had been set up in 1866 with financial support from Société Générale de Belgique. In 1908, when the Congo became a Belgian colony, Union Minière helped to form Compagnie Industrielle Union (CIU), which opened a copper smeltery and refinery at Oud-Turnhout in the Antwerp area. The core of the group's mining and metallurgical operations was formed at that time. Union Minière was responsible for the supply of ores and CIU specialised in copper metallurgy, while Vieille Montagne produced zinc. ACEC-Union Minière has since become the holding company of Société Générale de Belgique for the non-ferrous metals sector.

Soon after the First World War, a new development was the start of exploitation of mineral deposits in Zaire, then the Belgian Congo, which led to great progress in the metallurgy of copper, tin, cobalt, precious metals and other allied metals.

Immediately after the Second World War the shortage of raw materials had a serious effect on the Belgian industry, which experienced a difficult period. It was not until 1955 that the sector's total output again exceeded the 1938 level. The trend from 1955 in crude metals and semi-manufactures is as follows:

	Crude metals	Semi-manufactures
1955	469 300 tonnes	198 500 tonnes
1965	672 800 tonnes	371 400 tonnes
1975	757 900 tonnes	472 200 tonnes
1985	900 800 tonnes	683 400 tonnes
1990	1 024 100 tonnes	825 300 tonnes
1991	968 764 tonnes	785 311 tonnes

The production of aluminium semis also expanded in Belgium after the Second World War. Output, virtually nil before the war, rose from five tonnes in 1946 to 386 000 in 1989.

PROSPECTING AND EXPLORATION

Belgian mining legislation, like the French Mining Code, is directly derived from Napoleonic law (Law of 21 April 1810) and has remained very close to it, operating or exploration permits never having been introduced. It was codified by Royal Decree in 1919 (Lois Coordonnées) and subsequently amended in points of detail.

A distinction is made between deep mines, open-cast mines and quarries:

— the definition of *mines* in force takes the form of a listing of facilities and still corresponds to that of the Law of 1810;

— the term *minières* (open-cast mines) has been used since 1957 to include alluvial iron ore deposits, pyritic minerals suitable for processing to iron sulphate, aluminous minerals, dolomites and limestones used in steelmaking, plastic clays and brickmaking materials for large brickworks;

— *quarries* are also defined in the form of a list, which was also updated in 1957.

The production of materials falling under the heading of "minerals" is currently considered to be in the national economic interest. While minerals may not be extracted without permission, the state may grant mining concessions for such resources, even where there is opposition from the landowner.

However, minerals production is now limited to salt and sodium sulphate. It should also be pointed out that mining, as in 1810, remains a non-commercial operation not subject to licensing.

Minerals exploration may not be undertaken without the consent of the landowner or authorisation from the government (renewable after two years). No minerals exploration permit is exclusive.

Exploration may only be undertaken under the terms of a concession, which is held in perpetuity. The rights of the landowner are wide-ranging. He has preferential rights in the granting of concessions, subject to certain conditions. The state may nevertheless give preference to a person making a find or to a person requesting an extension.

Mines may not be sold or transferred, under any form whatsover, divided, let or leased without prior authorisation from the government.

The ending of mining concessions does not fall within the powers of the executive. It has to be decided by a court of law. A concession can only be revoked by judicial decision and the mine may only be returned to operation by a new grant of concession. Different rules, which are still in force, apply to the now exhausted bauxite mines.

MINING AND METALS PRODUCTION

Minerals taxation

The holders of mining concessions must pay land rights annually to the owners of the land, which are fixed (according to area) and proportional (1-3 per cent of the net profit). As a replacement for the proportional duty on mines, a tax on real profits and income has been introduced. Since 1991 the rate for companies is 39 per cent of taxable income.

Direct and indirect support

Belgian Government intervention in the mining and metallurgical industry normally takes the form of aids to enterprises in the sector which are encountering financial difficulties. The role of the regional governments is crucial in these types of action, which is usually taken on a case-by-case basis.

Moreover, under the provisions of regional development policy, Belgian firms may qualify for the following subsidies in certain regions of the country:

- interest rates subsidised to 5 per cent below market rates for five years on loans granted by government agencies to finance investment in fixed assets. The interest rate reductions cover up to 75 per cent of the investment and in certain cases the reduction may be increased to 7 per cent;
- if the investment is 50 per cent financed from internal funds the enterprise may receive, instead of an interest rate subsidy, a capital subsidy equivalent to the interest rate reduction which would have been available under the first heading;
- guarantees from the state or government financial institutions;
- exemption from the 1 per cent land tax for a period of five years;
- possibility of doubling the normal volume of depreciation for three succesive years on investments in machines, equipment and industrial plant;
- possibility of exemption from certain local taxes for five years (in particular, social charges and energy consumption taxes).

HORIZONTAL POLICIES

Environment

With regard to constraints imposed on mining operations and metallurgical plants in the interests of environmental protection, there are no regulations specific to such concerns in Belgium. They must comply with the rules applicable to all industrial enterprises in the country. There are, however, sectoral standards for the discharge of industrial effluent and special operating requirements for each industrial enterprise (emissions into the air, water and earth); these requirements generally depend on the location.

Since 8 August 1980, environmental regulations have fallen within the jurisdiction of each of the three regions of the country and are applied on a case-by-case basis. Only very general standards — such as those applicable to SO_2 emissions or industrial safety — are covered by national measures. In August 1980 the powers of the regions were widened, responsibility being devolved to them for both laying down and monitoring pollution control standards.

Waste management controls are exercised through the system of waste disposal consents. Waste disposal operations undertaken without prior authorisation are subject to fairly severe penalties. The regions reserve the right to impose a lien on the property of those who fail to pay the charge. The money will serve to build up funds to aid measures which may encourage or favour recycling.

The following examples of charges per tonne in the Flemish region for the main waste products from the mining and metal industry illustrate this approach:

- BF 30 per tonne for the dumping of waste on a single purpose tip authorised for mining waste;
- BF 80 per tonne for iron oxides arising in zinc production (jarosite and goethite);
- BF 100 per tonne for dumping on a random tip waste derived from the breaking up of cars and scrap metal objects;
- BF 200 per tonne for dumping waste on a tip authorised for inert materials;
- BF 300 per tonne for the incineration of waste in a furnace authorised for industrial waste and for the physico-chemical treatment of waste in installations authorised for the purpose;
- BF 350 per tonne for the collection of waste in the Flemish region to be dumped or incinerated outside the region;
- BF 500 per tonne for the destruction by incineration of toxic waste in an authorised plant; BF 400 per tonne for destruction other than by incineration and for the neutralisation and elimination of toxic waste, excluding the recycling and regeneration of recovered material;
- BF 900 per tonne for the dumping of industrial waste on a class I tip;
- BF 1 000 per tonne for the incineration of waste in a furnace not intended for the purpose;
- BF 2 000 per tonne for the disposal of waste on a tip not intended for the purpose;
- BF 3 000 per tonne for the incineration, neutralisation and elimination of toxic waste in unauthorised installations.

The same rates were adopted by the Walloon region.

In the mining and metal industry, an example of the effects of legislation is the conflict between the authorities and a number of companies in the sector which do not wish to be regarded as waste recovery operators. The effect of the new provisions is to make waste and refuse broadly equivalent, which is currently the subject of discussions with the Commission in Brussels. The conditions now being applied are much more draconian for non-ferrous metallurgy and may even threaten the economic viability of the companies, which specialise in the processing of complex raw materials and waste products.

At a time when their external sources of potentially toxic wastes are drying up because of the ban on international trade in this type of product, the cost of supplies from internal sources are rising substantially for many firms. Some of them are also having to absorb the higher costs of dumping waste from reprocessing secondary raw materials and of controlling lead and cadmium emissions more strictly. Such emissions are currently a key concern in the Belgian metal industry in the anti-pollution drive.

As for the other non-ferrous metals, there are no standard rules for the control of lead emissions to the atmosphere, which comes under the responsibility of the regional Environment Ministries. Limits are laid down individually for each plant and, in most cases, are geared to the TA Luft guidelines in Western Germany (Federal Republic of Germany before the unification of Germany). For lead dust emissions, the general limit is 10 to 20 mg/m^3. The regional authorities, which have held responsibility since 1987, base their guidelines on the proposals contained in Community directives.

With regard to the control of lead in industrial effluent, the limits in force since 1977 also vary in relation to the nature of the plant and the type of water used in the metallurgical processes. The limit is normally between 0.05 and 3 mg/litre.

The general field of pollution control measures also includes the draft European directive on the protection of workers against carcinogenic agents at places of work.

Also, a Belgian Intersectoral Committee on Cadmium was set up in 1988 grouping enterprises engaging in the production and primary processing of cadmium. This Committee plans to produce a review of knowledge relating to cadmium and the various questions raised by its use. An inventory of cadmium in Belgium and a White Paper on the toxicological and epidemiological evaluation of the metal and its compounds have been produced.

The Environment Minister for the Flemish Region has prepared a policy plan (MINA-PLAN 2000) for the period 1991-1995. Under this plan a body is to be set up to manage all aspects of the environment (water, land, waste, noise, etc.). Far-reaching changes will be made to the ministerial department so that it will be better able to carry out its task.

Research and development (R&D)

In 1982 the Belgian Government ceased to grant public aid for the promotion of R&D. On the other hand the Belgian metallurgical industry takes an active part in European R&D programmes on raw materials.

Trade policy

As a Member of the EC, Belgium applies the common external customs tariff, the section of which relating to the six main non-ferrous metals is summarised in the chapter dealing with the European Community.

CONSUMPTION/SUPPLY

Securing overseas supply

Belgium has always had close ties with Zaire, the main supplier of copper and cobalt to its metallurgical industry. Though supported by diplomatic arrangements, most activities in the field of supply are conducted by private Belgian companies, notably ACEC-Union Minière and its subsidiaries.

In the context of the progressive reduction of its mining base, notably as a result of the nationalisation of its Zairean assets at the end of the 1960s, Union Minière has taken measures designed to reduce its vulnerability on the supply side. These measures consist mainly in the signature of long-term supply contracts and in the expansion of recycling.

Recycling

Belgian plants have specialised in the processing of complex ores, residual products and wastes, thereby easing the pressure of supply constraints and promoting better use of by-products from the materials processed, in particular precious metals and metals arising in small quantities (gold, silver, platinum, palladium, rhodium, tantalum, niobium, antimony, selenium, tellurium, indium and bismuth).

MHO is a significant example among others, which thanks to a sustained policy of technical research, has both maintained its position as the top copper refiner in Europe and has become Europe's leading silver refiner, the world's leading refiner of germanium, rhodium and indium and the world's largest producer of special cobalt products. The recycling of waste from the electronics industry, rich in precious metals, is another focus of this company's activities.

In recent years, MHO has specialised in the hydro-metallurgical reprocessing of electrolysis residues, the recovery of impure zinc from alloy scrap and the recovery of zinc, copper and aluminium from scrap metal in various forms, such as car wrecks and domestic electrical appliances. Important quantities of non-ferrous metals are also recycled by other companies in this sector.

BIBLIOGRAPHY

Eurométaux (1989), *Conditions of Profitability of the EC Mining Industry*, Brussels, November.

International Lead and Zinc Study Group (1988), *Environmental and Health Controls on Lead*, Geneva, August.

Moniteur Belge, 30 December 1989.

Coordinated Laws on deep mines, open-cast mines and quarries.

Fédération belge des entreprises de métaux non ferreux (1990), *Statistical yearbook*.

CANADA

INTRODUCTION

Broadly defined, Canada's mining and metals industry comprises about 275 operating mines with their ancillary mills, 30 non-ferrous smelters and refineries, 59 primary iron and steel mills, and some 200 stone quarries and commercial scale sand and gravel pits scattered throughout the country. In the non-ferrous sector, Canada is a leading exporter of aluminium, magnesium, nickel, lead and zinc, and a significant producer of copper. This is consistent with the country's natural resource base. Canada depends wholly on foreign sources of bauxite and alumina for making aluminium. The Canadian mining and metals industry is almost exclusively controlled and operated by private enterprises and has significant levels of foreign ownership. The high dependence on export markets for five of the six metals can be readily discerned from the statistics in Table 2.3.

The provinces have responsibility for and jurisdiction over resources, including minerals, which lie within their boundaries. The federal government has jurisdiction over interprovincial and international trade, and matters affecting the national interest. The federal government also assumes the equivalent of provincial responsibilities over the Yukon, Northwest Territories and offshore territorial waters beyond provincial jurisdictions.

The federal government's non-ferrous policy is subsumed in the 1987 policy statement which defines its role towards the minerals and metals sector. The policy statement presented six broad objectives supportive of the exploration, development, production and utilisation of the nation's mineral and metal resources:

a) to provide a fair and balanced fiscal and regulatory framework;

b) to foster the development of the minerals and metals sector as a foundation for regional economic development;

c) to promote improved technological performance and increased international competitiveness in all facets of the industry;

d) to assist workers and communities affected by industrial adjustment;

e) to facilitate enhanced mineral and metal exports and access to new and traditional markets;

f) to promote timely and accurate economic, technical and scientific information required by the industry and by the federal and provincial governments, labour and the general public.

The statement goes on to say that the private sector is best placed to allocate resources among alternative activities and investments. Surviving and prospering in international mineral and metal markets require speed and flexibility, with decisions taken on the basis of sound commercial considerations. At the same time, the statement says, fundamental and ongoing changes world-wide in mineral and metal supply, demand and technology, cannot be altered by government. It concludes that in such a highly competitive world environment, the viability of the Canadian minerals and metals industry cannot be maintained through subsidies or other artificial support.

PROSPECTING AND EXPLORATION

Geological services

Essentially, all non-ferrous metals exploration, and indeed for all minerals, is conducted by the private sector. Basic geological mapping and the provision of geoscience data bases are shared by the provincial governments and the federal government. At the federal level, the Geological Survey of Canada (GSC) (a component of the Department of Energy, Mines and Resources — EMR) is responsible for the geoscience knowledge base of Canada including the offshore. The GSC provides information and advice relating to: planning, development and policy formulation; mineral exploration and assessment; oil and gas resource estimates; exploration and discovery of mineral and hydrocarbon resources; geological engineering and hazards; and logistic services for multidisciplinary scientific investigation in the high Arctic.

In recent years, the federal and provincial governments have co-ordinated some of their work by jointly funding Mineral Development Agreements (MDAs) to implement measures designed to stimulate mineral exploration and development by the private sector. Under the MDAs, funding is available for geoscience, technology and certain types of economic development activities. The geoscience programme is aimed at developing and improving the geoscience information base with which the industry can make decisions on exploration expenditures more effectively. The range of activities include surveys of metallic mineral deposits, regional geology, geochemistry and geophysics. Geoscience outputs are available from the GSC. The technology programme is designed for mining and processing activities and details are covered below. The economic development programme involves analytical studies to identify mineral development opportunities.

Also at the federal level, another component of EMR, the Survey, Mapping and Remote Sensing Sector (SMRSS), is responsible, inter alia, for national geodetic surveys, the preparation and distribution of topographic and geographic maps, and for the acquisition and compilation of remote sensing data. The SMRSS also manages property surveys of Canada Lands (Indian reserves, national parks and all Crown lands in the Territories and the offshore). The provincial governments assume responsibility for the property surveys of their respective territories.

The Mineral Policy Sector is responsible, inter alia, for the collection of exploration and mineral statistics and the maintenance of a number of data bases.

Lands availability

Most of Canada's land mass is public land, with surface and subsurface rights vested in the Crown, be they provincial or federal. Even in the populated regions of the country, the subsurface rights are vested, for the most part, in the Crown. Each province has its own mining laws and regulations concerning lands availability. Federal regulations control access in the Northwest Territories and in the Yukon. Exploration rights on public lands are usually obtained by staking claims and these rights are maintained by performing assessment work. Governments have prohibited exploration rights in national parks, conservation areas, areas for settlement of native claims and for migratory bird sanctuaries. Growing competition for access to land from competing economic interests as well as the perceptions and realities of the environmental impact of mining activities is of increasing concern to the Canadian mineral industry.

Canada has a land area of 9.97 million square kilometres. The percentage of Canada's land mass owned by natives will increase in the future, as more land claims are settled, but it is impossible to predict to what extent. None of the land controlled by natives is legally alienated from mineral production. Companies wanting to explore or develop on these lands must negotiate with the natives as they would with any landowner.

Currently, 1.39 per cent of Canada is devoted to national parks which are closed to any form of natural resource exploitation. The park system is to be expanded significantly by the year 2000 through the planned creation of 18 new parks, for the most part in the "mid-North" of the country. Federal bird and wildlife sanctuaries total 2.32 per cent of the land area. In some of these, resource exploitation is restricted but not altogether prohibited. Provincial parks approximately double the amount of land devoted to parks and the amount of parkland under provincial jurisdiction is expected to increase significantly. In the past, provincial governments have allowed resource development in parks but this policy is changing under the pressure of public opinion. Existing operations in provincial parks may continue but new activities are unlikely to be permitted. The share of total land unavailable for exploration and mining activities is 5.2 per cent, which consists of national parks and most provincial parks, as well as some special cases like the Yukon Game Sanctuary. Environmental groups in Canada believe that 12 per cent of Canada should be protected as wilderness areas, a goal which has attracted political support.

Mining claims and exploration permits

Authority for exploration permits and the issuance of mining claims in provincial territories fall within the purview of the respective provincial governments, usually the Department responsible for mines. On Canada Lands (those under the authority of the federal government), such authority resides in the Department of Indian Affairs and Northern Development. The issuance of mineral exploration permits or concessions is usually restricted to individuals or corporations having residential status in the province concerned. Such residential status requirements are not a significant restriction on Canadian or foreign owned corporations as such requirements are not difficult to establish.

Tax incentives

Firms or individuals that engage in mineral exploration are subject to federal income tax and provincial income and mining taxes. Similarly, each province and the federal government are able to set up specific mineral incentive programmes designed to suit their own needs. The following paragraphs describe current federal measures and the provincial incentive measures or programmes which are likely to materially affect exploration levels or patterns.

Federal taxation

All expenses incurred to determine the existence, location, extent or quality of a mineral resource in Canada are defined as Canadian Exploration Expenses (CEE) and can be written-off at up to 100 per cent against corporate income. A corporation whose principal business is mining must deduct CEE up to the amount of its income. Other taxpayers may deduct up to 100 per cent of CEE expenses against any income. Since many exploration firms do not have offsetting income and need private financing to carry out exploration work, they are able to issue what are referred to as "flow-through shares" to an investor who incurs CEE on behalf of the company. The attractiveness of these flow-through shares had been further enhanced by earned depletion allowances and by the Canadian Exploration incentives Program (CEIP). Both of these additional incentives were phased out in early 1990.

Provincial taxation

Some provinces have set up specific exploration incentive programmes designed to suit their own needs. For example, in Ontario, the Ontario Mineral Incentive Program (OMIP) provides grants of up to C$ 150 000 to cover 30 per cent of preliminary exploration costs. Also, the Ontario Prospectors Assistance Program (OPAP) is intended to encourage grass roots exploration by providing grubstake grants of up to C$ 10 000 per year to qualified individuals. In Quebec, flow-through share investors are entitled to an additional write-off of 25 per cent of underground exploration and 75 per cent of surface exploration expenditures financed by flow-through shares. Ontario and the Northwest Territories give a three-year tax exemption (with limits) to new mines and Manitoba exempts income from a new mine up to the amount of capital investment.

Direct participation

The federal government is not engaged directly through crown corporations (state entities) or indirectly through equity ownership in non-ferrous exploration activities, except for Cameco's gold-related operations. Cameco, owned by the governments of Saskatchewan (61.5 per cent) and Canada (38.5 per cent), is primarily a uranium mining and processing entity.

MINING AND METALS PRODUCTION

Mining

Infrastructure support

Canada's economic history has been dominated by natural resource development, which in turn has advanced hand-in-hand with regional and infrastructure development (roads, services, ports, schools, hospitals, etc.). This is particularly true in frontier areas, and non-ferrous mining operations sometimes influence these decisions. There is no programme as such designed to provide infrastructure support to the non-ferrous or any other sector of the mining industry. However, all levels of government on occasion make decisions about infrastructure.

Environment

Environmental considerations are as acute in Canada as anywhere in the industrial world and most industrial activities are scrutinised for their environmental impacts by federal or provincial agencies. Partly because of the visible effects on the environment of mining and smelting activities, these activities have long been regulated by provincial and federal governments and sometimes by both levels in a more or less co-ordinated manner. The following paragraphs contain only the major elements of the federal and provincial regulations and programmes affecting the Canadian mining and metals industry.

At the federal level, the Canadian Environmental Protection Act (CEPA), enacted in 1988, consolidated existing federal environmental statutes and provides for the powers to establish environmental guidelines for federal departments and agencies. The core of the Act relates to the identification, assessment, scheduling, and ultimate regulation of toxic substances. Through its responsibility (shared with the provinces) for fisheries, the federal Fisheries Act prohibits the discharge of substances into the fish habitat which are deleterious to fish or man's use of fish. The federal government has also significant jurisdiction in environmental matters under its powers of peace, order and good government. The Canadian Constitution makes no specific reference to legal jurisdiction over the environment. Consequently, jurisdiction over the environment is a shared responsibility, with governments relying on specific powers accorded to them under the Constitution to exercise authority related to the environment. The federal government exercises provincial-type responsibilities for mining and the environment in the Yukon and the Northwest Territories.

In 1984, the Environmental Assessment Review Process Guidelines Order (EARP) was approved by the Governor in Council. In 1989, the courts determined that EARP was a law of general application covering a broad spectrum of federal interests. In 1992, the Canadian Environmental Assessment Act (CEAA), which will replace the EARP Guidelines Order, received royal assent. CEAA was to be proclaimed in early 1993.

Canada signed the Basel Convention on 22 March 1989. Canada's Export and Import Regulations under the Canadian Environmental Protection Act (CEPA) will control the movement of hazardous wastes entering or leaving Canada. These regulations will also enable Canada to ratify the Basel Convention. Most trade in recyclable metals has been addressed in both the OECD Decision of 31 March 1992 and the Canada-USA Agreement on the Transboundary Movement of Hazardous Waste. Canada is currently in the process of ratifying the Basel Convention with the intent of participating fully in the first meeting of the contracting parties.

In 1985, the federal government and seven provincial governments established the Canadian Acid Rain Abatement Program (ARAP) to assist companies in meeting more stringent environmental standards. The programme is an important element in Canada's commitment to reduce sulphur dioxide emissions by half by 1994. In 1988, an acid mine drainage research and development programme was initiated to find ways of preventing sulphides (found naturally in some ores) from combining with oxygen and dissolving in water to form acid leachate which would enter the hydrological cycle. The programme is referred to as the Mine Environment Neutral Drainage (MEND) Program and is jointly funded by the federal government, five provinces and the industry.

According to the constitutional responsibilities of the provinces for land uses and exploitation of natural resources, each province applies environmental, health and safety, and labour regulations to mineral exploration and exploitation activities, such as mining, milling, smelting and refining, carried out on its territory. Environmental regulations relate to water pollutants, air pollutants, land rehabilitation, mine tailings, sulphur dioxide and nitrogen oxide emissions, effluent discharges and management of toxic waste. Health and safety regulations apply to a whole range of working activities and involve for example the quality of the working environment, exposure to noise, dust and fumes.

Minerals taxation

Federal taxation

The federal government imposes general taxes on fuels that apply equally to mining operations. Insofar as mining outputs are concerned, the federal government levies a 28 per cent corporate income tax, a capital tax of 0.2 per cent on the capital of large corporations and, in the territories, mining royalties and taxes.

For purposes of federal income taxes, a resource allowance deduction is permitted to recognise the non-deductibility of provincial mining taxes and royalties. The resource allowance is set at 25 per cent of resource profits, that is, operating profits less depreciation allowances but before deductions for exploration, interest and depletion.

Most capital assets are depreciated for tax purposes at rates approximating economic depreciation. However, a number of assets do receive accelerated write-offs for incentive purposes. Certain assets acquired for new mines and major expansions are eligible for accelerated write-offs to recognise the high risk of mining capital investments.

Taxpayers may write-off most mining assets on a 25 per cent declining balance basis. Assets for a new mine or major expansion are eligible for an additional write-off up to the income from the new mine or major expansion. The assets covered by this provision include mine buildings, machinery and motive equipment, processing assets (concentrator, smelter, refinery), and related social infrastructure such as roads and sewers.

Capital expenditures on research and development (R&D) may be written off immediately in the year incurred instead of being amortised over a period of years.

The cost of equipment purchased to control or limit air and water pollution can be depreciated 25 per cent in the first year, 50 per cent in the second year and 25 per cent in the third. This is available only for changes to a plant in operation before 1974.

The cost of certified machinery and equipment used for the generation of electricity or for the production or distribution of heat can be depreciated 25 per cent in the first year, 50 per cent in the second year and 25 per cent in the third.

Provincial taxation

The provinces impose provincial corporate income tax at a rate varying from 14.5 to 17 per cent and 6.33 per cent in Quebec, and mining taxes varying from 15 to 20 per cent based on income from mining computed at the mine mouth. Other significant levies include: provincial sales taxes and capital taxes, property taxes at the provincial and municipal levels and payroll taxes at the provincial and federal levels.

Metals production

Inputs pricing

In Canada, electricity is generated and distributed primarily by provincially regulated enterprises operating under monopolistic conditions, and is some cases offering incentives to certain large volume consumers, including the non-ferrous processing industry. The aluminium industry in Canada is located primarily in British Columbia and Quebec, provinces with abundant hydro-electric resources which enable the generation of low cost electricity. In Canada, electric utilities price electricity at the average cost of production, which is generally lower than the marginal cost of production. While marginal cost pricing has the advantage of achieving economic efficiency (i.e. closing the cost-price gap), it has not been adopted by utilities, mainly due to the complexity of the marginal cost pricing scheme. In addition, low-cost electricity has been seen in many provinces as an effective provincial economic development tool.

Direct and indirect support

There are no direct or indirect support programmes aimed at the non-ferrous sector. There are, however, generally applied labour adjustment programmes such as unemployment insurance, relocation and cost-shared job retraining programmes which become operative when mines or processing facilities close temporarily or permanently.

Regional development policy

In 1988, the federal government made some changes to its regional development policy by creating regional development agencies with mandates to undertake economic development measures that are tailored to the needs and priorities of their respective regions, i.e. the Atlantic Canada Opportunities Agency (ACOA), the Department of Western Economic Diversification (DWED), and a distinct organisation within the Department of Industry, Science and Technology Canada (DISTC) was assigned responsibility for regional development in Ontario and Quebec.

Each of the above three regional development organisations was given responsibility for: managing Economic and Regional Development Agreements (ERDAs) signed with the respective provincial governments and co-ordination of a number of subsidiary agreements (the Mineral Development Agreements, referred to in the section on "Prospecting and Exploration" are examples of ERDA subsidiary agreements); developing and implementing new programmes tailored to the needs of each region with an emphasis on small and medium-sized business; co-ordinating all federal regional development activities in its region; and monitoring and analysing trends in provincial economies. Most of the financial assistance provided to private enterprises through the organisations take the forms of repayable loans, loan guarantee programmes and grants for non-commercial projects. Although none of the programmes is directed at the non-ferrous sector, financial assistance is available for projects such as infrastructure which are compatible with regional development objectives and which, at the same time, may be of direct or indirect assistance to non-ferrous mining or processing activities.

Government ownership

The Canadian non-ferrous mining and metals industry is largely owned and operated by privately held companies. A number of foreign government-controlled companies are active in Canada. The exception, Cameco, (jointly owned by the governments of Canada and Saskatchewan), is explained above in the section on Prospecting and Exploration.

HORIZONTAL POLICIES

Foreign investment controls

Canadian government policy on foreign direct investment is centred in the Investment Canada Act (ICA) and in the amendments to the ICA contained in the Canada-United States Free Trade Agreement (FTA). The ICA, promulgated in 1985, replaced the former Foreign Investment Review Agency (FIRA) and its mandate was broadened to include the encouragement and facilitation of foreign investment in Canada. The Act provides for a review process of foreign investment above a minimum threshold: C$ 5 million in assets for direct acquisitions and C$ 50 million for indirect acquisitions. New or "greenfield" investments are no longer subject to the review process.

In reviewing transactions, Investment Canada assesses whether the acquisition would likely be of "net benefit" to Canada, taking into account various factors: the effect on employment; on productivity; on technological development; on international competitiveness; and considering the compatibility of the investment with national industrial, economic and cultural policies. Although the investment review procedure has delayed and sometimes modified investment, no proposed non-ferrous industry investment or take-over has been denied. With the exception of uranium, Canada does not have any restriction on foreign ownership in the minerals and metals sector. However, the federal government monitors the non-ferrous sector for significant changes in foreign ownership and control.

Under the FTA, the Canadian threshold for the review of US direct acquisitions was raised to C$150 million by 1992, and the review of US indirect acquisitions will be phased out over the same period. (An indirect acquisition would be when one US firm takes over a second US firm which has a Canadian subsidiary.)

Research and development

The Canada Center for Mineral and Energy Technology (CANMET), of the Department of Energy, Mines and Resources Canada, provides scientific research and technology development services with regard to mining, mineral processing, extraction and utilisation of minerals and metals. Its research programmes serve government responsibilities in areas of policy such as health and safety, protection of the environment, and economic and regional development. In addition to pure science, a major portion of the research and technology development related to minerals and metals is responsive to the needs of users and includes the latter's involvement on a full- or shared-cost basis.

The Province of Quebec maintains the Centre de recherche minière which conducts research in support of Quebec's mineral industry. The Provinces of Ontario, Saskatchewan and Alberta have research establishments which, inter alia, are available for research needs of their respective mining industries. They operate mainly on a cost-recovery basis and on occasion do some pure research.

Trade policy

Border measures

Under the Export and Import Permits Act, the federal government has the legislative authority to:

a) ensure that any action taken to promote the further processing in Canada of a natural resource that is produced in Canada is not rendered ineffective by reason of the unrestricted exportation of that natural resource;

b) limit or keep under surveillance the export of any raw or processed material that is produced in Canada in circumstances of surplus supply and depressed prices and that is not a produce of agriculture.

That legislative authority — in the form of export embargoes or export taxes — has however never been applied to non-ferrous trade.

At the provincial level, some provinces (Manitoba, Ontario, Quebec and New Brunswick) have legislation designed to increase the likelihood that natural resources extracted from their territories will be processed in those provinces. The province of Ontario has legislation requiring companies to obtain authorisation to export unprocessed metal bearing materials. For example, INCO Limited and Falconbridge Limited must obtain authorisation to export nickel matte and nickel oxide to the United Kingdom and Norway. However, these authorisations are usually for ten-year periods and have never been denied. Thus, intermediate products are exported out of these provinces and even out of Canada.

Canada-United States Free Trade Agreement

The Canada-United States Free Trade Agreement (FTA) came into force on 1 January 1989. The Agreement provides for liberalisation in all sectors of bilateral trade, including the mineral and metal sectors and also includes binding commitments on trade in services, business travel and investment. The wide scope of the Agreement is indicated from the outset in the agreed objectives, which are to:

a) eliminate barriers to trade in goods and services between the territories of the parties;

b) facilitate conditions of fair competition within the free trade area;

c) liberalise significantly conditions for investment within this free trade area;

d) establish effective procedures for the joint administration of the Agreement and the resolution of disputes;

e) lay the foundation for further bilateral and multilateral co-operation to expand and enhance the benefits of the Agreement.

The FTA provides that each side grant national treatment to each other's investors in respect of new laws or regulations, and that export, local content, local sourcing or import substitution requirements cannot be imposed.

Concerning import and export measures, the Agreement confirmed the rights and obligations provided for in the GATT and mentioned that both parties understand that any other form of quantitative restriction or minimum export-price is prohibited. As no such restrictions currently affect the bilateral non-ferrous trade, full implementation of the FTA will not bring changes to this sector. The comprehensive trade Agreement provides, also, for the gradual elimination of barriers to trade between the two countries. For non-ferrous trade, this means that all bilateral import tariffs will be eliminated, some were eliminated on 1 January 1989 and others will be eliminated following a five or a ten year period. In response to representations from industry, the governments of the two countries have been discussing which items might benefit from an acceleration of the tariff reductions.

North American Free Trade Agreement

On 12 August 1992, Canada, Mexico and the United States reached an agreement in principle on a North American Free Trade Agreement (NAFTA). The Agreement is expected to be implemented in the three countries' respective domestic legislations by 1994. A NAFTA will promote the efficient use of minerals and metals in the North American region as a whole, encouraging economies of scale and promoting intra-industry specialisation. As a result, Canadian firms should benefit from lower costs and higher productivity, helping them compete more effectively in world markets.

CONSUMPTION/SUPPLY

Securing overseas supply

There are no governmental measures specifically designed to give financial assistance for exploration of non-ferrous metals (nor indeed other minerals) in territories outside Canada, nor to acquiring minerals from abroad. Canada has a wide range of expertise, both in the private and public sectors that are involved in mineral resource development and management abroad. These range from geological mapping and electronic data interpretation, to remote sensing techniques, mining technology and mineral processing. These services are provided on a commercial basis via consultants, special services firms and Canadian-based multinational mining companies. Some of these services are also provided through Canada's foreign aid programme and on occasion the Export Development Corporation provides export financing for mining equipment and machinery to be used in foreign projects.

Canada firmly believes that international consultations and comprehensive, up-to-date information on mineral and metal production, trade and consumption are crucial to sound decision-making on non-ferrous metals supply and procurement. In this regard, Canada has traditionally supported international producer-consumer dialogue and information exchange. Canada was a founding member of the International Lead and Zinc Study Group and the International Nickel Study Group. Canada has recently joined the International Copper Study Group.

Stockpiles

Canada has no government stockpile for non-ferrous metals.

Recycling

Metal recycling reconciles economic growth and environmental preservation. It successfully completes the life cycle of metals and minerals from exploration to extraction through to use and re-use. Metal recycling lessens the stress on landfills, recovers potentially hazardous substances, reduces the demand on the mineral resource base and, at the same time, saves energy. The Government of Canada is committed to the development of new and improved technologies that promote recycling.

Canadian trade (imports and exports) of metal recyclables averages 4 million tonnes per year. The value of this annual trade is in the order of C$ 2 billion. Non-restrictive trade policies for metal recyclables remain a high Canadian priority. In co-operation with other OECD countries, Canada was an active participant in the negotiations resulting in the OECD Council Decision of 31 March 1992 which is designed to facilitate the unhampered trade of recyclable materials. This agreement, consistent with Article 11 of the Basel Convention, clearly distinguishes between recycling activities and disposal activities. Incorporated into this Decision is a mechanism which provides for varying degrees of transboundary movement control depending upon the hazard level posed by the individual recyclable item. The Government of Canada is committed to transboundary controls on the movements of hazardous recyclables which strike a proper balance in the nature and severity of controls to ensure both that recycling is encouraged and the environment is preserved.

FRANCE

INTRODUCTION

France is a major producer of non-ferrous metals, in particular alumina, aluminium, lead and zinc. Plants producing aluminium, copper and nickel semi-manufactures are also a French speciality. Some traditional mining activities are in decline however. This applies to bauxite, in which production in 1989 was a quarter of what it had been ten years previously, and lead, output of which is only one sixteenth of what it was at the end of the 1970s. Iron, potassium and uranium mines continue to operate. On the other hand, reserves of nickel in New Caledonia are estimated to be the second greatest in the world, not far behind those of the Commonwealth of Independent States (CIS); current nickel ore production ranks third in the world.

France's policy on non-ferrous metals was mapped out in 1975 in the wake of the first oil crisis. Its two main objectives are:

— to reduce dependence on external sources of supply by the development of mining and metallurgical activities and the promotion of recycling;

— to ensure the availability of ores and metals consistent with France's needs and to reduce the foreign exchange costs of supply from outside sources.

One of the specific features of this policy has been the key role assigned to the Bureau de recherches géologiques et minières (BRGM), which has developed its activities along four main lines:

— inventorisation of French territory with a view to identifying new metallogenetic provinces and to localising new deposits;

— identification of new natural resources abroad as a prime instrument of French co-operation policy in the mining field;

— exploration on a commercial basis with a view to setting up a portfolio of economically viable deposits, which in certain cases could be exploited in conjunction with other partners;

— supplying of services in the field of "geological engineering" in the wider sense to different types of operator: governments, international organisations, local authorities, industrial concerns.

Since 1980 French policy has been influenced by several factors, chief among them being the overcapacity which has arisen on markets for several metals and decreased vulnerability in respect of certain metals (including copper).

A more detailed analysis of the national supply picture has been undertaken on a product by product basis, leading in some cases to preference being given to vulnerability responses other than integration upstream to mine level: namely, the development of new technologies, substitution by other metals and precautionary stockpiling, the main objectives being to reduce the cost of imports, strengthen the competitive position of French enterprises in the sector and protect metallurgical activity in the country.

This note describes the body of precautionary and support measures adopted in times of difficulty for France, including the panoply of aids and incentives introduced in January 1975 following the first oil shock. Laws and regulations are rarely repealed, instead only some of them are applied depending on circumstances at the time (notably budgetary trade-offs reflecting government priorities, and successive Finance Acts). After 1983, markets were liberalised and in practice, the initiative lay once again with the users and processors of non-ferrous metals. The role of French operators in the sector has gradually emerged as a decisive factor in the national supply process. The government thus began to devote special attention to measures needed to place the financial, fiscal and legal environment of French firms on the same footing as that of competitor countries. At the same time it began to give closer consideration to links downstream in the chain, going to, the aid of major operators who had been particularly hard hit, such as Péchiney, Société métallurgique le Nickel (SLN) and Peñarroya. But its basic policy of creating a competitive environment for its industries, together with support of a small, but active, precautionary stockpile, public programmes of research, including recycling and raw materials conservation, and international co-operation particularly in countries of potential supply was reconfirmed by the council of ministers in june 1990.

PROSPECTING AND EXPLORATION

The heavy involvement of the state, including the use of political and economic measures, is one of the main elements in France's minerals policy where non-ferrous metals are concerned. One of the main aspects of the policy here consists in the implementation by government agencies of systematic research and exploration programmes on French home territory and abroad conducted.

Geological services

Created in 1958, the Bureau de recherches géologiques et minières comprises all government institutions and agencies responsible for the conduct of mineral exploration in France and its Overseas Territories.

The BRGM enjoys financial support from the French Government in the form of budgetary appropriations and contracts to enable it to implement a mineral exploration programme and to play an industrial role alongside other mining and metallurgical enterprises in France. An industrial and commercial public undertaking, the BRGM does not enjoy the same measure of financial autonomy as an ordinary enterprise. In particular, it is not allowed to borrow from banks unless specifically authorised to do so by its State Supervisor ("Contrôleur d'État") or make trading profits on more than roughly half its turnover (that part which comes neither from subsidies nor from special government orders).

Since 1968 the BRGM has also been responsible for the Service géologique français. Within the framework of the role assigned to it, the BRGM in 1975 decided to carry out five-year programmes for the inventorisation of mineral resources in metropolitan France. These involve a general effort of prospection centred on deposit indicators followed by extensive dissemination of the results obtained. In 1988 the project to produce a detailed geological map of French territory on a scale of 1/50 000 was more than two thirds completed. A mineralogical inventory of Guyane was also launched in 1975, financed by the state[1].

Lands availability

Environmental concerns have prompted changes in French mining law to accommodate the interests stemming from such concerns. However, the creation of nature parks and reserves entailed a risk that large portions of territory might be rendered sterile, with bans on prospection for, and the extraction of, minerals. For that reason, an agreement was reached between the Ministry of Industry and Research and the Ministry for the Quality of Life to ensure that mineral exploration and mining projects should never be prohibited outright on grounds of conservation or environmental protection. Each individual project would, where appropriate, be subjected to special examination. It is in this context that impact studies were introduced in France.

Exploration permits and concessions

The principle applied is that of national sovereignty over underground resources. Prospection concessions are granted by the Direction des matières premières et du sous-sol (DMPSS — Directorate for Raw Materials and Undergound Resources) of the Ministry of Industry.

Prospection activities are not taxed and may be undertaken on the simple consent of the owner of the land (or subject to compensation, the amount of which is fixed by the courts). The DMPSS may, however, grant exclusive prospection rights without the prior consent of the landowner. Prospection permits are issued for an initial term of three years, renewable for a further period of up to three years. The extension may be made subject to payment of a certain sum of costs negotiated in advance. If the holder of an exclusive prospection permit finds an economically viable deposit, he is entitled to a mining concession.

The law is designed to provide greater encouragement for the development of deposits by the issue of concessions for a maximum duration of 50 years, by strengthening the conditions for withdrawing concessions which are not being used and by widening the field of basic information on underground resources which must be communicated to the administration. All transfers of mining rights must be authorised by the government. Mining rights may be revoked by the state in the event of non-payment of taxes for a period exceeding two years, cessation of mining activity or pursuit of an activity contrary to public interest.

Tax incentives

The exploration activities of mining and metallurgical enterprises qualify for special tax relief. Thus, for example, reserves for expenses or future losses may be set up, provided such expenses are assigned to the financial year in question and are clearly defined and deducted.

A coordinated scheme for the encouragement of mineral exploration was devised within the framework of the Plan métaux (Metals Plan), which set the terms of French Government financial aid for specific projects. This scheme was discontinued in 1986.

Direct participation

During the 1970s the BRGM accounted on average for almost two thirds of French mineral exploration activities (excluding uranium), which ranks it among the world's leading mineral exploration concerns. With regard to other research activities, a small number of operators were responsible for the great majority of the work carried out. Prominent among these companies were Peñarroya, Cogema and the subsidiaries of the Péchiney, Total and ERAP-Elf/Aquitaine Groups. During the 1980s, the relative share of work undertaken by the BRGM in global mineral exploration activities increased still further.

MINING AND METALS PRODUCTION

Mining

Infrastructure support

Albeit infrequent, mention might nevertheless be made of the construction or modernisation of certain regional infrastructures in the Nord/Pas-de-Calais region, in particular engineering work carried out on the river network which links the area with Belgium, the Netherlands and Germany. This made it possible to operate large capacity barges on the route, which will reduce transport costs for imported ores, such as zinc concentrates consumed by the Noyelles-Godault and Auby plants.

Environment

French policy on energy and pollution control have had effects on mining activities on French territory. Thus, for example, in the 1970s the mining industry was obliged to sign contracts with the Agence française pour la maîtrise de l'énergie, undertaking within a certain number of years to secure improvements in the specific consumption of energy per tonne of ore produced. Moreover, for any mining project to be authorised, it must first be subjected to an environmental impact study.

There are no hard and fast rules for the control of non-ferrous metal emissions to the atmosphere, which falls within the jurisdiction of the regional authorities. For dust emissions in general, the maximum permissible levels vary between 5 and 30 mg/m^3. The standards applied for the maximum quantity of lead permitted in the air of communities located in the vicinity of metallurgical plants are those of the EEC (2 mg/m^3 as an annual average), and supervision is exercised by the local authorities. With regard to lead in industrial effluent, the regional controls in force limit discharges to 1 mg/litre.

The limits for lead at places of work are controlled by the Ministry of Social Affairs and Employment. Maximum levels were set in 1986 at 0.15 mg/m^3 air (for an average of 40 hours). The limit for lead in the blood of workers is supervised by the same Ministry. There is no precise standard for this parameter, but it is normally between 70 and 80 µg/100 ml. Occupational physicians also monitor levels of amino-uaevulinic acid, which must not exceed 20 mg/litre.

Minerals taxation

There is a tax on the activities of the minerals industry which is payable by mining companies and the holders of mining concessions. For non-ferrous metals, the tax is FF 0.40 per hectare. Duties on mining activity, collected by the local authorities, also have to be paid in respect of each ore on the basis of quantities produced. The rates are fixed each year by interministerial decrees.

Mining companies are exempted from trade tax. Non-ferrous metals companies are also entitled to deferment of VAT payment. VAT deferment is also applicable to scrap from non-ferrous metals and their alloys.

French mining enterprises also qualify for the deposit depletion provision [provision pour reconstitution de gisement (PRG)]. Enterprises producing solid minerals of interest to the French economy are permitted to set up a reserve not exceeding 50 per cent of their profits and 15 per cent of the value of sales of extracted products. The reserve must be devoted within a period of five years to the financing of research or development work within certain geographic areas specified by decree as priority zones; outside such zones the reserve may in principle only be used to finance half the work undertaken. This reserve, which applies to most minerals[2], nevertheless carries some restrictions with regard to the assessment basis applied and conditions of employment.

On the other hand, social security charges on wages and salaries in French mines are significantly higher than those in other industries. In metal mines they amount to rates generally ranging between 80 to 90 per cent of the wage bill[3].

Metals production

Inputs

Following the two oil crises, the French electricity industry launched a very ambitious investment programme geared mainly to nuclear energy. In 1988 the electricity produced from thermonuclear sources already accounted for 70 per cent of total French production. By the end of that year the nuclear power inventory operated by Electricité de France (EDF) consisted of 55 units rated for a total of 52 GW.

The increase in demand, however, sustained by conversions to electricity in industry, was considerably lower than projected by forecasts. To counter this situation, EDF launched an adjustment policy along three main lines: increase exports, downgrade capital expenditure plans and support the establishment on French territory of high energy consumption industries. The most controversial issue is however the one concerning energy tariffs offered in attempt to maintain in operation big energy consuming industries and to attract new investments to these areas.

This is the case of the tariffs paid by the big aluminium producer Péchiney to EDF for the energy consumed in aluminium production. Under a first agreement, Péchiney took a share in the capital of a nuclear power plant, in return for a commitment by EDF to supply 2 billion kWh per year for a 25-year period, agreeing in principle to increase amounts taken in two stages to 5 billion kWh.

The capital investment allowed Péchiney to pay a price for the electricity close to the operating costs of the nuclear plant (about 10 centimes per kWh, or about one half of its total cost), as compared to the 16 centimes per kWh paid until then. Péchiney's stake does not involve a specific nuclear plant but a hypothetical one to give the aluminium producer complete flexibility in its electricity supplies.

A new long-term electricity supply agreement with EDF was negotiated in 1985, involving 3 billion kWh per annum for a ten year period. Under this agreement, Péchiney did not pay cash but issued to EDF FF 2 billion worth of non-voting securities ("titres participatifs"). The two agreements cover more than 50 per cent of the group's basic electricity needs for its aluminium and metal fabrication business in France.

This agreement was replaced in 1986 with a cheaper contract covering the next five to eight years. All of the advantages of the 1985 contract were concentrated in the shorter version, which reduced costs from over 20 centimes per kWh (price paid for the electricity by big consumers) to little more than the 11.2 centimes in the 25 year 1983 contract.

Finally, in 1989, Péchiney made a deal with EDF for the joint development of a new aluminium smelter in France, to be located in Dunkirk. Under the deal, EDF will supply the smelter with electricity for 25 years, at an initial cost of around 6 centimes per kWh rising by 10 per cent each year for the three years between from 1997 to 1999[4]. In spite of these rises, some analysts think that in the future it will be very difficult for EDF to keep Péchiney's power supply under its marginal production costs.

Direct and indirect support

The French Government has set up a special agency, DATAR (Délégation à l'aménagement du territoire et à l'action régionale), to coordinate all the activities of the various government agencies in the field of regional development. For the purpose of the subsidies extended by this body, France is divided into four zones (from A to D), three of which (A, B and C) qualify for special conditions. These include cash subsidies covering up to 25 per cent of the sum invested by enterprises creating or developing activities in Zone A and in special regions of Zones B and C.

Although a number of conditions are attached to these subsidies, their calculation is closely geared to the number of jobs created by the investment. There is a limit of FF 25 000 per person employed per loan and a ceiling of 25 per cent of the investment to which the subsidy relates. Given the capital intensive nature of the activity, it was considered that the effect of this measure was limited to about 1 per cent for the aluminium industry.

Guyane also qualifies for a tax incentive scheme for mining and metallurgical projects. Economic feasibility studies for the exploitation of mineral deposits are financed by the state.

Where New Caledonia is concerned, there is an industrial development policy to optimise the exploitation of the territory's nickel deposits. This policy falls into two main parts:

- As far as SLN was concerned, French Government policy, applied by the Ministry of Industry, consisted in the acquisition by the state-owned firm, Elf-Aquitaine, in 1974 of 50 per cent of the shares of the mining company. The oil concern thereby contributed to a major investment programme which was to increase the capacity of the Doniambo metallurgical plant from 70 to 90 kt/year of nickel contained in mattes and ferro-nickels. The measures also included the commissioning of a high-purity nickel plant at Le Havre. The total investment in the two units was about FF 1 billion. In 1985 fresh state intervention involved the financial restructuring of the company through the intermediary of the state holding company, ERAP, which became a majority shareholder. The state action has been justified by the fact that nickel represents by far the most important economic activity in New Caledonia.
- In connection with the development of a new mining and metallurgical project, following the abandonment by the Patiño Group of the project it had embarked on in the north of the island in 1973, the French Government decided to promote a new project in that area. To achieve this, the BRGM acquired control of 90 per cent of COFREMI, which

holds major deposits in the north of the island. This project has not yet been implemented.

Loans are available for all industrial activities (except for minerals exploration) from Crédit National, a state bank specialising in development, and from the Compagnie de développement régional. The interest rates on these loans have not been subsidised since 1986. However, any investment project contributing to an increase in economic activity or to regional development is eligible for several types of loan, some of them at interest rates below market levels, subject to approval which is given on a case-by-case basis.

After the nationalisations of 1981 (see below) and against the background of a severe recession in the world mining and metallurgical industry, the government came to the aid of major operators who had been particularly hard hit: Péchiney, SLN and Peñarroya. Thus FF 7 billion were injected into Péchiney in the form of share capital or virtual share capital. SLN and Peñarroya also received substantial public support.

Government ownership

After the change in government in 1981, the non-ferrous metals industry was for all practical purposes nationalised in France. This intervention was achieved in different ways: directly, by nationalization pure and simple, as in the case of Péchiney, or indirectly. The latter heading covers state acquisition of control resulting from the nationalization of companies having subsidiaries in the mining and metallurgical sector (e.g. Imetal-Peñarroya) and acquisition of control following financial restructuring (e.g. SLN).

Péchiney: The Law of 11 February 1982 resulted in the nationalization of PUK, a group which at the time comprised the following divisions: steel and titanium (Ugine Aciers), aluminium (Aluminium Péchiney), chemicals (Produits Chimiques Ugine Kuhlmann), mining and electrometallurgy (Sogerem, Sofrem), nuclear power and research, special products (Métaux spéciaux) and copper processing (Tréfimétaux).

By decision of the Ministry of Industry, Ugine Aciers in 1982 came under the control of Sacilor. The chemicals division was quickly dismantled, as in 1983 the government decided to restructure the French chemicals industry. Finally, the other metallurgical activities remained within the Group which changed its name from PUK to Péchiney.

Imetal-Peñarroya: The French State became the majority shareholder in Imetal in 1981, following the nationalization of Banque Rothschild and Cie. Financière de Suez and the acquisition of shares in the ERAP Group and Compagnie générale des matières nucléaires (COGEMA). In 1983, given the financial difficulties experienced by the Group, which arose in particular from the effects of the non-ferrous metals crisis on the Peñarroya subsidiary, the French Government decided to strengthen its control over Imetal. ERAP (which also controls SLN) and COGEMA thus became the main shareholders and acquired responsibility for the strategic development of Imetal.

Taking advantage of the recovery in non-ferrous metal prices in 1987, Imetal sold most of the Peñarroya shares owned by the Group on the stock market. After the creation in 1988 of the company, Metaleurop, by a merger between Peñarroya and some non-ferrous metal concerns of the German Group, Preussag, Imetal reduced its interest in that subsidiary to less than 15 per cent.

SLN: In view of the crisis on the nickel market, Imetal in 1974 sold 50 per cent of the shares of SLN to SNEA/Elf-Aquitaine, which assisted SLN in its investment programme during the second half of the 1970s. This effort however was not enough to prevent SLN from accumulating substantial losses: over FF 2.2 billion between 1977 and 1983.

In 1983, a new nickel crisis led to a reconfiguration of the share structure and the industrial and financial restructuring of SLN and to the acquisition by ERAP of a 70 per cent interest in the firm, which made it possible to take SLN out of the consolidated accounts of Imetal. Imetal and SNEA retained 15 per cent each. Finally in 1985, SLN was split up into two companies: "Société métallurgique le Nickel-SLN" retained its industrial assets in New Caledonia and became a subsidiary of a new company named "Eramet-SLN", which holds the Sandouville refinery, markets all the finished products and provides technical assistance to SLN.

In addition, most other companies directly or indirectly engaging in exploratory activity are under some degree of state control, on the lines of the BRGM, (COGEMA) and Société nationale Elf-Aquitaine (SNEA) (although the latter two do not share the same legal status, nor does the state participate in them for the same reasons).

HORIZONTAL POLICIES

Foreign investment controls

There is no restriction on the activity of foreign companies provided their subsidiaries are registered as French companies. However, the government reserves a right of veto on the transfer of certain activities to foreign operators if they are deemed to be of strategic importance to the nation.

This right, normally exercised by the Ministry of Industry or the Ministry of Finance, has been tacitly invoked on several occasions, for example in connection with mining development in New Caledonia. More recently, the Ministry of Industry backed the refusal of Péchiney to sell the Noguères aluminium plant — the closure of which was scheduled for 1991 — to the Swiss group of the trader, Marc Rich. This form of control has become much more flexible, however, as European integration progresses.

Taxation

In 1991, France set a corporate tax rate on non-distributed profits of 34 per cent. This rate rises to 42 per cent on dividends paid to shareholders and other beneficiaries. Main deductions are the following:

- accelerated depreciation: immediately upon completion of new construction, the qualifying businesses may depreciate 5 per cent of its cost (50 per cent for R & D buildings, anti-pollution equipment and equipment used to economise on primary materials);
- withholding tax: 25 per cent or the reduced tax treaty rate (between 5 and 15 per cent);
- "provisions pour fluctuation des cours" (PFC): firms principally engaged in the processing of specified international commodities, the prices of which are closely linked to their quotations on international markets, may establish tax-free reserves to cover price fluctuations of such commodities. The ceiling on such reserves is calculated in relation to a basic stock of the relevant commodities;
- "provisions pour hausse des prix": may also be established with respect to price increases of any raw materials or supplies in stock, other than international commodities, for which the replacement cost has increased by more than 10 per cent within the period of two fiscal years. Only the price increase in excess of such 10 per cent may be carried to the special reserve. This tax exemption is only temporary and the amounts carried to this reserve must be reincorporated into income within six years.

Specific tax provisions are applicable in France's Overseas Departments and Territories (DOM-TOMs). When new companies are created (or new company activities begun) which employ at least five new persons, the fist ten years' income is exempt from the French tax. Prior approval is required.

Research and development (R&D)

Grants administered by the Ministère de l'industrie et de la recherche are made available to all scientific and technical research activities which promote the development of new techniques. Up to 50 per cent of the cost of a project may be obtained, depending on the probability of success, and the grant must be refunded in case of success. Aid is also available for initial demonstration plants and for industrial investments which will achieve savings on imported raw materials. Most of this assistance has been directed toward work on the substitution of copper in a number of applications. This aid has been granted on a case-by-case basis.

Aid to research and development (R&D): the government offers a 50 per cent tax reduction (up to FF 5 million per firm) on the following types of research and development expenditures applicable to the mining industry:

- the construction cost of buildings used solely for scientific and technical research, including prototypes and pilot installations;

- the cost of equity shares in government-approved research companies and organizations;
- the payroll costs for research;
- other operating expenses related to research and development activities up to a limit of 55 per cent of the aforementioned payroll costs;
- expenses associated with obtaining and maintaining patents.

This type of aid was also used in the context of specific R&D programs, such as:

- aid to the development of new technical processes for special operations, such as aluminium and nickel in the 1970s;
- upgrading of mineral resources programme, aimed at increasing research in the area of earth sciences.

Supplementing its exploration efforts, the BRGM lends scientific support to mining operations in France (as well as in foreign countries) to develop ore exploration, ore analysis and ore assessment techniques. In most cases, this work receives financial support from the Ministry of Industry and the European Commission. BRGM's scientific support to mining operations extends to prospecting for raw materials on the ocean floor where the company is active in the development of new beneficiation processes.

Trade policy and marketing

The action of purchasing groups enjoying government support is designed to increase negotiating strength, to promote a better knowledge of world markets and to facilitate the diversification of suppliers. In France such action has extended to coal, asbestos and copper, in the latter case through the Groupement d'importation et de répartition des métaux (GIRM). Dating back to the Second World War, the GIRM still undertakes copper purchases to meet the needs of private firms in France.

On the other hand, the lack of flexibility in exchange rate controls until 1987 slowed down the development of trade in ores and metals because it prevented arbitrage transactions on foreign markets. Finally, the price controls in force until 1975 constituted the measure which undoubtedly had the most adverse effect on production of those minerals and metals subject to controls (aluminium, ferro-alloys, sulphur, potassium).

As a Member of the EEC, France applies the common external customs tariff, the section of which relating to the six main non-ferrous metals is summarised in the Chapter dealing with the European Community. Working towards improving the rapidity and efficiency of the Community's anti-dumping procedure is also an announced objective of French policy, reconfirmed by the Council of Ministers in June 1990.

CONSUMPTION/SUPPLY

Securing overseas supply

Two historical factors have had a decisive influence on the course followed by French supply policy since the beginning of the century: on the one hand, the tradition of state intervention in the economy; on the other hand, the colonial past which has favoured the ventures of French mining operators outside Europe, seeking security of supply through a policy of maintaining a presence at points of production and exerting control over resources.

French authorities have pursued a general policy designed to minimize the vulnerability of the country to outside influences, particularly after 1975. However, this strategy of security through control has become a great deal more sophisticated and is currently achieved by the development of the human, technical and financial presence of French operators in the producer countries, whatever the precise contractual arrangements for their presence.

The system of international co-operation which has thus evolved within the framework of mutual benefit agreements covering, amongst other things, state guarantees for investments by French companies abroad, technical assistance for mineral exploration and development and public funding for the training of specialised technical staff in mining countries.

The basic idea was always to put together an "integrated package" involving the exchange, on the basis of bilateral technical assistance, of French capital and services for mineral products.

In this context, the role played by the BRGM in the implementation of this French supply programme has been absolutely crucial. Indeed some important tasks have been entrusted to the Bureau in the fields of mineral resource surveying, co-operation policy and intervention abroad. In addition some incentive schemes have also been applied directly to operators in conjunction with specific programmes, the best known example being the Plan cuivre.

Support of exploration

The main measures of support for mineral exploration activities abroad are as follows:

a) Since 1966, the BRGM has been contributing directly to the expansion and diversification of ore supply sources. The BRGM is active in some forty foreign countries, with three main spheres of operation: the Middle East, Central and West Africa, and Southern Europe (in particular, the Iberian Peninsula).

b) Availability of the depletion allowance for deposits abroad since 1972.

c) The Plan cuivre (Copper Plan), followed by the Plan métaux (Metals Plan) in the 1980s; set up in 1973 at the level of the Ministry of Industry, it comprises a Committee with the following tasks:

— to draw up mineral exploration programmes, programmes for participation in the exploitation of or in projects for the exploitation of copper deposits and in copper scrap reprocessing operations;

— to make selections from projects presented by firms;

— to fix the amount and terms of state financial aid on the basis of partial aid repayable in the event of success[5]. Loans were normally granted up to a ceiling equivalent to 50 per cent of the cost of the project approved. Generally speaking they were repaid at the rate of 120 per cent in the event of success.

The Plan cuivre made it possible to avoid an almost complete cessation of copper prospection activities by French operators and was instrumental in the launching of exploration activities in the following countries: France, Portugal, the Philippines, Saudi Arabia, Jordan and Zaire. The Plan cuivre led to the discovery by the BRGM and Peñarroya of the Neves Corvo deposit in Portugal. However, the mining rights for this deposit were sold in 1984, and the activities pursued under the Plan were suspended in 1986.

d) The special metals programme: set up in 1975 to enable the French industry to occupy some key positions for the future. It concentrated on titanium and zirconium, but preparatory measures also extended to lithium and tantalum. Projects were covered by specialisation agreements with German firms.

e) Programme on ocean seabed manganese nodules to provide a focal point for the work of the Centre national d'exploitation des océans (CNEXO), the Commissariat à l'énergie atomique (CEA) and individual firms.

Assistance to foreign producers

French co-operation in the field of mining is characterised by a heavy concentration of activities in the French-speaking African countries:

a) Participation in geological and mineralogical surveys and pre-investment studies in conjunction with support to companies and services involved in geological surveying and prospection activities in the producing countries. This work, for which France provides all or part of the funding through the Fonds d'action of the Ministry for Co-operation, is usually entrusted to the BRGM[6].

b) Training of specialised staff:

— Credit extended by the Fonds d'aide et de coopération (FAC).

— Creation in 1975 of the Centre d'études supérieures des matières premières (CESMAT) on the initiative of the Ministry of Industry and Research by the mining colleges, the mining companies and the BRGM. It serves as a structure for coordination between training establishments, providing training for overseas engineers both in France and abroad (by the secondment of French teachers).

c) Capital aid: the Caisse centrale de coopération économique (CCCE) may part-finance the acquisition by producer states of shares in the capital of joint operating companies set up on their territory. It operates mainly in French-speaking Africa, including Zaire. In this way, mining projects were financed in the second half of the 1970s in Niger, Senegal, Mauritania and Togo. The CCCE may also finance infrastructure projects in the mining countries of French-speaking Africa.

d) DIE (Développement industriel à l'étranger) loans, administered by Crédit National and used to finance two mining projects in Zaire.

e) Technical assistance to the mining countries provided by the BRGM under bilateral co-operation arrangements and financed through the French Government technical aid programme. This programme covers exchanges of information, technology transfers, consultancy services, human resource development and economic agreements under which the costs of exploration and the results can be shared.

Tax incentives

General tax concessions encourage French minerals and metals companies to operate abroad. Firms who sell or render services abroad and generate income that is taxable in France, and for which they grant medium-term credit, may place up to 10 per cent of the credit amount in a tax-deductible reserve without having to prove any actual credit risk.

French companies that establish sales, market research or information facilities abroad, directly or through a share of at least 10 per cent in a foreign company, may create a tax-free reserve which must be declared to French tax authorities. The reserve may be equal to the losses incurred in the first five years of operation up to the amount of capital invested during the period if the investment is within the EEC; or equal to the amount of capital invested during the first five years of operation in a non-EEC country. In the cases of investments in developing countries, French companies may create a tax-free reserve equal to 50 per cent of the capital invested during the first five years of operation. The reserves are subject to prior approval by French tax authorities and the reserve must be restored to taxable income in five equal fractions from the sixth to tenth years of operation.

Insurance to domestic companies investing abroad

The system of guarantees for which French firms are eligible in respect of their investments abroad was set up in 1972, initially only covering the French

Franc area. Subsequently the system was extended to countries having entered into a bilateral agreement. There are two different guarantee systems:

— the COFACE (Compagnie française d'assurance pour le commerce extérieur) rules are applied when the investment generates a satisfactory flow of French exports;
— the BFCE (Banque française du commerce extérieur) also offers loans and guarantees against the political risks of investing abroad. It is more flexible because it does not impose the use of French equipment; however it does take into account the significance of the project French supply requirements.

The system of guarantees for investments abroad has the following characteristics:

— only risks of war and expropriation can be covered;
— the guarantee is limited geographically to the French Franc area and to a small number of countries outside it;
— it covers the initial investment but not reinvested or distributed profits;
— compensation is paid for up to 75 per cent of the loss sustained but this may rise to 90 per cent in certain cases.

Direct intervention in projects abroad

French State involvement in this area began with action to group within appropriate structures French enterprises concerned with mining development abroad (mining enterprises, consumer industries; engineering consultancies, equipment manufacturers, financial agencies). In this context the company SOCOMINES (Société de coopération minière et industriel) was set up in 1975, with the participation of the BRGM (50 per cent), Usinor, Peñarroya, Sofremines and Sofresid, at the time of the nationalisation by Mauritania of the French iron ore producer, MIFERMA.

The prospection activities of the BRGM are conducted through a specialised subsidiary, SEREM (Société pour l'étude, recherche et exploration de minerais), while COFRAMINES (Compagnie française des mines) accounts for a sizeable portion of the group's mining portfolio. The objective assigned to COFRAMINES, with respect to the activities of both the BRGM and other French mining operators, was in particular to provide an industrial link downstream in the chain from the development of deposits discovered in the context of bilateral co-operation.

In addition to its public service role and its functions in the field of exploration and technological research therefore, the BRGM pursues an industrial activity financed in part by public money, particularly in French-speaking Africa (FAC credits). Thus, for example, the Bureau has received from the government the financial resources necessary to acquire shares in the Société Minière de Tenke-Fugurume (Zairean copper). The BRGM has also participated in the development of gold mines in Canada and Burkina Faso and controls holdings in phosphate mines in Senegal (Compagnie sénégalaise des phosphates de Taïba), manganese mines in Gabon (Comilog) and chromium mines in New Caledonia.

Stockpiles

A general security stockpile was set up in 1975 with the aim of providing, within ten year, two months of secure supply. It was not until 1980, however, that this policy got underway and, in practice, the results were more modest as its role was limited to emergency stockpiling to ride out crises.

Management is provided by the Groupement d'importation des métaux (GIRM). Financing is provided by lines entered annually in the budget of the Ministry of Industry and Research. The products, volumes and funding of the stockpile are strictly confidential. It seems, however, that economic criteria have taken precedence over strictly strategic criteria in the specification of the list of ores. In June 1990, French authorities reconfirmed that the stockpile would be managed according to the following principles: coverage limited to defined number of minerals; release in times of a major crisis; access for stategic civil industries (eg. electronics) as well as defense industries; et periodic updating of crisis scenarios by the Caisse française des matières premières (CFMP).

Recycling

The efforts of the government in this field are devoted in part to directly supporting the action of the BRGM under materials savings programmes. The latter are geared to securing material savings by way of adjustments to products at the design stage or to production processes. In parallel with this, the government has stimulated the recovery of waste and by-products from mineral and metallurgical production. Two national agencies are responsible for the coordination of recycling policy — ANRED (National agency for recovery and elimination of waste) and AFME (French association for the control of energy). AFME and ANVAR (National agency for valorisation of research) are also responsible for the promotion of measures designed to encourage raw material savings and substitution.

Among the measures developed in this context are:

- the improvement of recovery chains, e.g the setting up of a chemical residues exchange;
- sectoral action under the Plan cuivre, involving the development of waste reprocessing facilities in France and a manufacturing technology making it possible to recycle production waste without prior refinement;
- the treatment of urban waste using technical processes developed by the BRGM.

NOTES AND REFERENCES

1. FF 630 million were spent from 1975 to 1984 on the Inventaire Minier, FF 510 million of this in Metropolitan France. The main targets were: lead, zinc, tungsten, antimony, copper, tin, silver, barite, bauxite, fluorine.
2. In particular, aluminium, antimony, arsenic, beryllium, bismuth, boron, chromium, gold, lead, lithium, manganese, molybdenum, nickel, platinum, pyrites, rare earths, silver, tin, titanium, tungsten, vanadium, zinc and zirconium.
3. For example, accommodation allowances for retired staff are equivalent to 5 per cent of wages.
4. This rise was not contained in the original proposal and was added after an investigation by the European Commission, prompted by protests from Péchiney's competitors. *Power in Europe*, December 7, 1989.
5. FF 375 million were spent from 1973 to 1984. The principal beneficiaries of the loans were the BRGM, Elf-Acquitaine, Imetal-Peñarroya, Thomson-Brandt, Gogerem-Péchiney and Paribas.
6. Examples: Bolivia, Peru and Sudan.

BIBLIOGRAPHY

BOISSON, Pierre (1986), "Raw Materials Report" *Mission du Plan*, August.

BRGM, *Annual Report*, several years.

CHAZAN, Willy (1988), *French Mineral Exploration (1973-82)*, TILTON, J., EGGERT, R.G. and LANDSBERG, H. (editors): *World Mineral Exploration, Trends and Economic Issues*, Resources for the Future, Washington, D.C.

Eurométaux (1989), *Conditions of Profitability of the EC Mining Industry*, Bruxelles, November.

IMAUVEN, Claude (1991), "La politique française des matières premières", *Annales des Mines*, February, pp. 5-7.

Observatoire des Matières Premières (1983), *Stockpiling of raw materials in the industrialised countries*, Note Technique no. 21, OMP, Paris.

GERMANY

Taking recent developments and changes in Germany into account, an analysis of the mining and non-ferrous metals sector is best approached by examining the two former parts in order to obtain the most accurate picture of the German industry as a whole. The study is therefore divided into two parts: the non-ferrous metals industry in Western Germany (Federal Republic of Germany before the unification of Germany); and that in the new Federal Länder (the German Democratic Republic before unification).

PART I: THE NON-FERROUS METALS INDUSTRY IN WESTERN GERMANY

INTRODUCTION

Due to a limited mineral resource basis, only small quantities of lead and zinc concentrates have in recent years been extracted from domestic mineral reserves. Despite its limited domestic supply of non-ferrous concentrates, Western Germany has nevertheless a significant non-ferrous metals production capacity which relies on imported raw materials. To secure supplies of minerals, Western Germany is pursuing international co-operation with geological services of other countries and up to 1990 supported selected exploration minerals projects abroad.

In the light of depleted reserves, the mining of zinc and lead ores from the Grund and Meggen mines came to an end in the first half of 1992. The Meggen pyrites and zinc and lead ore mine had already reduced output in 1989 because of difficulties in disposing of pyrite waste oxides. In 1989, production of crude ores from both mines totalled around 979 000 tonnes, compared with 1.1 million tonnes in the previous year. The recoverable metal content amounted to 61 300 tonnes. Employment in the non-ferrous ore mines felt slightly from 772 in 1988 to 712 in 1990.

In Western Germany, about a third of the approximately 350 000 tonnes of refined lead output in 1988 was obtained from concentrates, whereas two thirds came from recycled materials.

PROSPECTING AND EXPLORATION

Geological services

The central institution in Western Germany for the geological sector is the Bundesanstalt für Geowissenschaften und Rohstoffe (BGR) in Hannover. This body, under the authority of the Federal Economics Ministry, advises on all aspects of the earth sciences and the raw materials industry. One of its key tasks is to assist in securing supplies of minerals and energy for Western Germany. In this context, it advises German industry and carries out prospecting projects on Federal territory and, more importantly, abroad.

Publications of the BGR include "Survey of supply and demand for unprocessed minerals", "Country reports on the minerals industry" and it also publishes geological maps of Western Germany and collaborates with international organisations in publishing international geological maps.

The BGR also contributes to geo-technical safety measures, deals with the geological aspects of environmental protection and carries out methodological and instrumental development work in the earth sciences. It co-operates actively in international earth sciences with foreign geological institutions.

The Information Centre on Minerals Extraction, Earth Sciences and Water Management (GEOFIZ) in the BGR is a central information service covering the mining of minerals, the earth sciences and water management. GEOFIZ receives payment from private subscribers for its services, including direct access databanks and commissioned research.

Reflecting the federal structure of Western Germany, there are Land Geological Offices which serve individual Länder and sometimes neighbouring Länder as well. These bodies are under the administrative and technical supervision of the Land Government or of a department. Their responsibilities include carrying out geological surveys and assessing all mineral reserves and deposits.

The Land Office for Soil Research in Hannover, Lower Saxony, houses the Joint Earth Sciences Mission. This institution, acting on behalf of the Land Geological Offices and the BGR, carries out projects in geophysics, geology of hydrocarbons and geochemistry, and the publication of findings of the whole Federal territory, including the German continental shelf.

Exploration permits and mining claims

The Federal Mines Act of 1980, the key piece of mining legislation, applies to prospection and exploration, extraction and preparation of mineral deposits. Non-ferrous metallic ores are among the mineral deposits which are not the property of the landowner. To locate deposits and extract minerals it is necessary to have a mining licence issued by the State. The Federal Mines Act distinguishes among the following types of mining licences: permits for prospection and exploration in a three-dimensional field; consents for the extraction of mineral deposits in a single field; and mine ownership. In granting exploration and mining rights, no restriction is placed on foreigners.

The Chief Office of Mines ("Landesoberbergamt"), or in its absence in a particular Land, the Office of Mines ("Landesbergamt"), are responsible for issuing the various permits and consents for exploration and mining rights. A permit for prospecting and exploration is issued for a defined area following the submission of a programme of work describing the nature, scale, purpose and time-scale of the prospection activities. The permit is temporary and subject to renewal. The holder of the permit must pay annual site taxes.

A consent covers all the associated activities which precede, accompany and follow extraction, such as opening up the mining area, pit ventilation and tipping. Upon application, the permit holder will be given a consent, if there is prospect for a mining operation to be economic.

Mine ownership confers wider rights and powers than a consent; it is entered in the land register and is counted as property against which loans may be granted (the equivalent of land title).

Lands availability

Exploration and mining activities are restricted in principle in such areas as national parks and wild-life reserves. Exploration or mining activities may exceptionally be allowed in such areas, but the burden of proof is on the firm to show that no emissions or any other damage will be caused by the planned activities. In 1989, the wildlife reserves (land and water) in Western Germany covered an area of around 343 800 hectares or 1.38 per cent of the territory. Since then, more than 200 wildlife reserves per year have been added. The four national parks occupy a further area of 558 800 hectares. In addition, conservation areas are designed for purposes of recreation and tourism.

MINING AND METALS PRODUCTION

Infrastructure support

The level of transport costs is crucial to the non-ferrous metals industry and its competitive position. In setting their rates, the state-run German Federal Railways offer competitive rates to the mining sector, among others, in order to enhance their own competitive status vis-à-vis competing carriers (lorries and internal waterways).

As regards the Economic Community's internal market, the system of State-fixed freight and transport rates is moving towards the elimination of restrictions on market access. One development is that, from 1 January 1989, transport has been chargeable on the basis of recommended prices. Rates for marginal freight on the Federal Railways and in internal waterway transport allow for flexible pricing.

Mining authorities

Supervision of mining activities is the responsibility of the Länder. A Chief Office of Mines supervises the mining offices, which in turn are under the authority of the Economics Ministry. The Federal Mines Act of 1980 established a national committee of experts on mining, on which representatives of the Länder mining authorities sit. This Committee advises the Ministry on matters of mining technology and mine safety. It also takes a position on draft mining regulations.

The mining and dressing of mineral ores are subject to the supervision of the Offices of mines. This also applies to the construction of mine installations and to all measures which are directly associated with mining processes. The main tasks of the mining authorities include risk prevention, the protection of mine sites and the legal procedures connected with mining permits. Their responsibilities also include implementing other statutes, such as labour protection, environmental protection, and landscape and area planning. On the basis of mining regulations, a mining entrepreneur must draw up plans for the proposed site and submit these to the mining authority for approval.

Mine safety and social insurance

The most important task of the mining authorities is their responsibility for mine safety. Under this aspect the mining authorities supervise mining and its impact, from prospecting through to the closure of mines.

The Mining Employer's Association, a federal public organisation, is the insurance agency in Western Germany for accident insurance in mining. Together with the health insurance and pension schemes, it is the core of social insurance in the mining sector. The Mining Employer's Association also maintains a research institute for occupational medicine, an institute for research on dangerous substances, two mine rescue stations and a mining test centre which carries out research on explosives.

The German mining industry has set up and developed its own educational system since the beginning of the 19th century. The educational opportunities and training courses range from training for technicians to courses for qualified engineers.

Taxation

The taxation of mining activities is governed by the general tax law. Mining firms are subject to corporation, industrial and property taxes. The corporation tax is set at 50 per cent of taxable income on retained (accumulated) profit and 36 per cent on distributed profit. German mining enterprises are also subject to a trade earnings tax.

There are special tax provisions for mining operations:

a) Expenditures for prospection and exploration are accounted as operating costs and deductible for tax purposes. Losses may be carried forward. As to exploration activities abroad, such costs are deductible for tax purposes.

b) Expenditures on reclamation measures are accounted as operating costs. Reserve funds may be established to finance prescribed reclamation measures.

c) Personnel employed underground receive a tax allowance of DM 10 for every shift worked underground.

d) Under the 1990 Tax Reform Act, special deductions allowed for certain types of investment in coal and ore mining were abolished. Also abolished were the general relief provisions for environmental protection and for research and development. Tax allowances for variations in prices and exchange rates, not confined to the mining industry, were terminated at the beginning of 1990.

HORIZONTAL POLICIES

Environment

The protection of the environment is governed by mining, industrial and water law, and a number of technical regulations:

— Act on environmental liability;
— Act on the protection of nature;
— Act on the water economy;
— Act on charges for disposal of waste water in watercourses.

The protection of nature requires, among other things, that mine installations and pit heaps be adapted to landscaping requirements. Competing claims for land use are co-ordinated through area and district planning. Federal planning laws, district planning laws and area development plans are the decisive factors. When mining is finished, and to some extent also while underway, reclamation measures must be adopted, under the supervision of the mining authorities.

The Federal Emissions Protection Act of 1974 contains rules on clean air and noise control. It offers a range of instruments for these purposes, including approval procedures, clean air plans and emission registers. The Technical Direction on Clean Air (TA Luft) sets maximum pollution levels and other requirements for facilities which require planning permission. The TA Luft is one of the standard-setting authorities for approvals under the Federal Emissions Protection Act. Another authority relevant to facilities subject to planning permission is the Technical Direction on Noise Control (TA Lärm). It sets standard emission levels and works to limit sources of noise.

The Federal Water Management Act also governs the use and drainage of water from mines. The Offices of Mines, which are also responsible for controlling water disposal, carry out annual water checks together with the State water and waste disposal authorities and other boards and authorities.

The Act on Environmental Impact Assessment came into force on 1 August 1990. This Act contains umbrella regulations for testing the environmental impact of certain public and private projects. In the case of mining projects, the assessment of environmental effects is governed by the provisions of the Federal Mines Act, amended in 1990 to incorporate the EEC Council Directive on environmental impact assessment. The process of approving mining projects is supplemented by a planning procedure which takes special account of the particular features of mining. Another regulation on environmental impact assessment for mining projects prescribes which mining projects are subject to tests and must undergo the planning procedure before obtaining approval. The regulation stipulates the documents to be submitted for the purpose of environmental assessment.

The Federal Emissions Protection Act, September 1990, covers atmospheric protection, and sets out the practical and legal precautions to be taken when approving projects.

Part I of the Technical Directions for Special Wastes sets out the requirements for storage, treatment and burning of special wastes. The accompanying regulations (defining wastes and residues and monitoring wastes and residues) came into force on 1 October 1990. Part 2 of the Technical Directions for Special Wastes, governing the requirements for the surface and underground storage of special wastes, also of importance for the non-ferrous metals industry, was submitted to the Federal Parliament for approval.

Research and development

The federal government promotes research in mining especially through the Labour and Technology Programme (AuT), but which is not sector specific. The Programme seeks innovative approaches to labour and technology to avoid or reduce excessive human strain or hazard. The measures adopted range from technical and organisational solutions to those involving occupational training and health. Approximately DM 7.6 million were made available for the programme in 1989. The projects financed from these funds were preponderantly in other areas than non-ferrous mining.

As regards the promotion of technology and innovation, the Federal Ministry for Research and Technology cut back resources in 1989 for the raw materials sector. Only DM 5.6 million was allotted to the mineral sector. This category included projects such as improving the processing of low-grade and complex ores and the production of metals. The Ministry also allocated about DM 126 million to the category of materials research. Special emphasis was placed on projects aimed at improving the properties of materials and developing new and especially non-metallic or inorganic materials, as well as projects on recycling and materials saving and substitution. In this category also, materials based on non-ferrous metals played a minor role.

Trade policy and marketing

As a Member of the EEC, Western Germany applies the common external customs tariff, the section of which relating to the six non-ferrous metals is summarised in the chapter dealing with the European Community.

The export of investment and manufacturing plant and equipment and associated services to developing countries can be assisted by an export financing programme. This programme is based on resources from the European Recovery Programme (ERP Special Funds) and capital market funds from the Credit Institute for Reconstruction (KfW). Credits from the KfW/ERP export financing programme are intended exclusively for exports to developing countries.

The Federal Finance Ministry may offer guarantees or pledges for export transactions to help exporters, thereby underwriting credit risks. There are several measures: manufacturers' risk guarantees and pledges; export guarantees and pledges; and finance credit guarantees and pledges. The exporter must also pay a consideration for guarantees and pledges, and must provide his own cover against accidental loss. Applications are processed by the Hermes Credit Insurance Company and the Trustee Company (Treuarbeit). The decision whether to grant a guarantee or pledge lies with an interministerial committee convened by the Federal Economics Ministry. Tied loans are used to finance exports by German firms. They are granted directly to the foreign customer as customer credits.

CONSUMPTION/SUPPLY

Securing overseas supply

Raw materials procurement policy

In view of Western Germany's high import dependency on non-ferrous metals and other raw materials, raw materials supply policy is oriented to supporting the economy through: diversifying channels of supply; developing process technologies which save raw materials; and in stepping up research and development on recycling.

The following foreign policy initiatives are designed to support the above: providing legal and political guarantees; working towards free trade in raw materials; developing co-operation on the basis of partnership between industrial countries and minerals-exporting developing countries; and developing international co-operation with geological services and relevant research institutes.

Promoting exploration and investment abroad

The federal government's exploration promotion programme came to an end in 1990. This programme, administered by the Federal Economics Ministry, supported selected minerals projects by German firms of exploration and feasibility studies both at home and abroad. Firms established in Western Germany were able to apply for funds, repayable on special terms, to supplement their own expenditure. The aim of this measure was to secure supplies of minerals by reducing exploration risks. Such funds were to be repaid if the ventures were successful. In 1989, DM 18 million were allocated under this programme and DM 11 million in 1990. However, non-ferrous projects were not emphasised.

Instruments used to promote foreign investment are not geared to specific branches of industry, but the non-ferrous metals industry can benefit from these instruments. For instance, Western Germany supports investments under certain conditions by German mining and metals firms to secure supplies of raw materials, especially projects in developing countries. The range of supporting instruments includes: capital investment guarantees; financial aids; and tax measures.

The Federal Ministry for Economic Co-operation promotes investments by small and medium-sized German firms in developing countries. These involve long-term loans under relatively favourable conditions. Credits may be sought for investment promotion on a pro rata basis, or to finance preparatory studies. The Credit Institute for Reconstruction is responsible for processing applications, but the German Finance Company for investment in developing countries (DEG) may also be, if this latter is involved.

The Federal Republic has concluded bilateral agreements on the promotion and mutual protection of capital investment (investment promotion agreements) with about 60 States, primarily developing countries. The main provisions of these investment promotion agreements cover the treatment of investors, most-favoured-nation status, protection against expropriation or nationalisation, the free transfer of capital and profits from capital funds, and international arbitration. Investment promotion agreements are the basis for granting a Federal guarantee for capital investment abroad.

Stockpiles

No State reserves of non-ferrous ores or metals are maintained.

PART II: THE NON-FERROUS METALS INDUSTRY IN THE NEW FEDERAL LÄNDER

INTRODUCTION

On 3 October 1990 the eastern federal Länder — the former German Democratic Republic (GDR) — united with Western Germany. The former GDR was a State trading country. This centrally-run type of economy cannot be compared with the economy in Western Germany. The new Federal Länder are undergoing a comprehensive restructuring process which is affecting the non-ferrous mineral and metals sector.

In the former GDR, the mining sector was able to meet much of the domestic demand for minerals, in particular for brown coal, potash and tin. A priority consideration was to save foreign currency. This self-sufficiency policy however was achieved by operating low grade deposits at high operating costs, several time higher than world prices in the cases of copper and tin.

Against an uncertain economic future, the copper mine in the Sangerhausen district was closed in August 1990; it produced a copper concentrate of 40 per cent copper content. In early 1990, the Mansfeld AG plant employed about 6 200 workers and in 1991 an estimated 1 400 were employed on disassembly work. In Hettstedt also operated by Mansfeld AG, the copper and silver works employed about 1 500 workers in 1990 and a copper and aluminium semis factory employed about 6 800 workers.

There are plans to establish a modern secondary metallurgical plant in the former Mansfeld combine and a new lead works, making use of existing installations at the former VEB mining and foundry combine of Freiberg, drawing on West German participation via the Metallgesellschaft AG. By 1992/93 the capacity of the lead works could double to 45 000 tonnes a year.

On 3 October 1990, the tin mine at Ehrenfriedersdorf was closed because it had become uneconomic. It had been producing about 300 000 tonnes of crude ore a year, with a tin content of between 0.15 and 0.2 per cent. Of around 560 employees in 1990 about 200 are redeployed on conservation work. The tin mine at Altenberg also proved uneconomic and closed in March 1991. Until 1989 it had been producing about 1 million tonnes of ore a year, graded at 0.3 per cent Sn.

Following the cessation of all non-ferrous mining in the new federal Länder, the remaining non-ferrous processing and metallurgical plants have had to find alternative sources of supplies and restructure. In 1991, the three processing plants listed in Table 3.5.2 were still operating. Recycling will doubtless play a decisive role in future.

The cessation of copper and tin mining entailed substantial conservation, environmental and social costs. It is estimated that DM 188 million were required in 1990-91 for the conservation of the copper mines and DM 207 million for the tin mines. Social provisions for redundant workers are estimated to have cost DM 83 million in 1990. Estimated environmental costs related to ecological damages inherited are significant.

PROSPECTING AND GEOLOGICAL SERVICES

Prospecting for minerals, ground water and geothermal resources, and environmental and geological investigations were the responsibility of 20 State institutions. In 1990, their financial position deteriorated seriously. The State prospection funds were drastically reduced, being cut from DM 91 million in 1989 to DM 42 million in 1990. Research funds for crude minerals were also reduced by half.

By analogy with Western Germany, the new federal Länder mining authorities are placed under the aegis of the Länder and Land geological offices which partly incorporate existing institutions. The responsibility of the Bundesanstalt für Geowissenschaften und Rohstoffe and the Joint Earth Sciences Mission, of the Land Office for Minerals Prospection, now extends to the new Federal Länder as well.

MINING LAW

With the Unification Treaty, the application of the Federal Mines Act was extended to the new Federal Länder since 3 October 1990. However for a transitional period, certain other provisions of the former mining law continue to apply. For example, prospection for minerals is now governed entirely in accordance with the Federal Mines Act, but previous rules apply for the quarrying of stone and soils. In other areas, statutes and regulations are being taken over or adapted, as necessary.

INFRASTRUCTURE

The transport infrastructure in the new federal Länder is in extremely poor shape. This is especially true of roads and railways, and is bound to have an impact on the transport of non-ferrous metallurgy products. A priority target, which includes immediate measures to improve the transport infrastructure, is the reconstruction and extension of the historic east-west axes. Municipal and district councils, chambers and associations of commerce and industry may apply for additional funds for infrastructure developments, as long as they are necessary for the economy.

ECONOMIC ASSISTANCE FOR THE NEW FEDERAL LÄNDER

In the light of the historic transition from a planned to a market economy, and the special difficulties associated with it, both the European Community and the federal government are granting temporary aids and promotional subsidies. The Federation, Länder, EC and the large trade associations are working together on these aid measures. They will be implemented for a limited time and will be of particular importance for small and new businesses. Incentives apply equally to foreign investors.

The following are of relevance to the non-ferrous metals industry:

a) loans to establish businesses (by private persons, small and medium-sized firms) and for investment in environmental projects;

b) loans from the Credit Institute for Reconstruction (KfW) for the formation, consolidation of enterprises and for environmental improvements;

c) loans from the Deutsche Ausgleichsbank für Investititionen for the establishment and consolidation of enterprises and for innovation;

d) measures under the Community programmes for the improvement of regional economic structures, to promote industrial investment and regional infrastructures.

BIBLIOGRAPHY

Anger, G., Deutscher Auslandsbergbau — unternehmerische Aktivitäten und verbandliche Gemeinschaftsaufgaben. In: Jahrbuch Bergbau, Öl und Gas, Elektrizität, Chemie 1991. Essen 1990, pages 1-36.

Bergbau-Berufsgenossenschaft: Jahresbericht Bergbau-Berufsgenossenschaft 1989. Bochum, August 1990.

Bundesanstalt für Geowissenschaften und Rohstoffe: Tätigkeitsbericht 1987/88. Hannover, April 1989.

Bundesanstalt für Geowissenschaften und Rohstoffe: Jahresbericht zur Rohstoffsituation 1988. Hannover, June 1989.

Bundesministerium für Forschung und Technologie: Faktenberich 1990 zum Bundesbericht Forschung 1988. Bonn, March 1990.

Bundesministerium für Forschung und Technologie: BMFT Förderungskatalog 1989. Bonn, May 1990.

Bundesministerium für Wirtschaft: Wirtschaftliche Förderung in den neuen Bundesländern. Bonn, May 1991.

Bundesministerium für Wirtschaft/Bergbehörden der Länder: Der Bergbau in der Bundesrepublik Deutschland 1989 — Bergwirtschaft und Statistik. Trans Tech Publications, Clausthal-Zellerfeld 1990.

Bundesministerium für wirtschaftliche Zusammenarbeit. Deutsche Unternehmen und Entwicklungsländer. Bonn, February 1987.

Bundesministerium für wirtschaftliche Zusammenarbeit: Journalisten-Handbuch Entwicklungspolitik '90/91. Bonn 1990.

Eurométaux: Conditions of Profitability of the EC Mining Industry, Bruxelles, November 1989.

Gocht, W., Wirtschaftsgeologie und Rohstoffpolitik. Springer Verlag, Berlin/Heidelberg/New York/Tokyo 1983.

Gute Aussichten, trotz schwieriger Rahmenbedingungen für die deutsche NE-Metallindustrie. Metall-Interview mit J. Stegmann, Präsident der Wirtschaftsvereinigung Metalle e. V. In: METALL *45* (1991), No. 1, pp. 52-55

Jahrbuch Bergbau, Öl und Gas, Elektrizität, Chemie 1991. Verlag Glückauf GmbH, Essen 1990.

Kleinherne, H. und G. Schmidt: Die Umweltverträglichkeitsprüfund von Projekten des Steinhohlenbergbaus und der Energiewirtschaft — Chancen und Risiken.

In: Jahrbuch Bergbau, Öl und Gas, Elektrizität, Chemie 1989/90. Essen, pp. 1-16.

von Mässenhaussen, H.-U.: "Auswirkungen der Umverträglichkeitsprüfung und des Planfeststellungsverfahrens auf den deutschen Steinkohlenbergbau" Speech before the Congress on Hardcoal mining in Bochum on 24th January 1991.

Mining Annual Review 1990. London 1990.

Statistisches Amt der DDR: Statistisches Jahrbuch der Industrie der Deutschen Demokratischen Republik 1990. Berlin, September 1990.

Statistisches Bundesamt: Statistisches Jahrbuch 1989 für das Ausland. Wiesbaden, December 1989.

Statistisches Bundesamt: Aussenhandel, Fachserie 7, Reihe 2, Aussenhandel nach Waren und Ländern (Spezialhandel), December and year 1989. Metzler-Poeschel, Stuttgart, June 1990.

Statistische Bundesamt: Statistisches Jahrbuch 1990 für die Bundesrepublik Deutschland. Wiesbaden, September 1990.

US Bureau of Mines: Minerals Yearbook, 1987, Vol. III. Washington, 1989.

Wirtschaftsvereinigung Bergbau: Das Bergbau Handbuch, Verlag Glücklauf GmbH, Essen 1983.

Wirtschaftsvereinigung Bergbau: Zum Bergbau in der DDR — ein vorläufiger Überblick. In-house Studies, April 1990.

GREECE

INTRODUCTION

Greece is Western Europe's biggest bauxite and nickel producer and the only chromite producer in the EEC. The country also produces lead and zinc and will soon start producing gold. Greece is a net exporter of bauxite, alumina and aluminium metal. It is also an important source of supply for Europe of bauxite, non-ferrous metals, ferro-alloys and industrial minerals. Larco, for the production of ferro-nickel, and Aluminium de Grèce are the two biggest base metals firms.

In 1985, the entire metallurgical sector accounted for about 4.3 per cent of Greek GDP and that year the value of exports was in the order of 18 per cent of the country's total exports. The value of mining production is divided between energy minerals (56 per cent), metallic ores (16 per cent) and industrial minerals (28 per cent).

As from 1981, government policy has been to increase state control over the economy and, in particular, the mining industry. As a result, by 1987 over half the Greek mining companies were state-controlled. That same year the Government started to gradually change its policy towards the liberalisation of the sector. The main factors behind this new orientation were the decline in the competitiveness of the sector, falling private investment and EC pressure.

The government thus made efforts to promote foreign investment, notably in alumina and bauxite production, but this did not have the success hoped for. Similarly, in order to meet the undertakings agreed to for the single European market after 1992 and to halt the decline in the mining sector, the government stopped restrictive practices (notably the ban on dismissals) and abolished the subsidies allowed for administration and for the co-management of workers.

There are a substantial number of enterprises of all sizes, both public and private, in the sector. Among the biggest are Aluminium de Grèce (a 60 per cent subsidiary of the Péchiney group), the Bodossakis foundation (for lead and zinc concentrates), the Eliopoulos Kyriacopoulos group (barytes, bentonite and bauxite, operating the Bauxite Parnasse Mining Co.) and Hellenic Industrial and Mining Investment Co. (HIMIC).

The state enterprises are developing their activities in the production of bauxite, chromite and lead in the non-ferrous sector. Among the most important are Larco (for ferro-nickel), Emmel, Fimisco and Hellenic Ferro-Alloys.

During its presidency of the EEC in 1988, Greece endeavoured to promote the definition and implementation of a European mining policy. This resulted in the adoption of the Mining Resolution of June 1989, the broad lines of which are described in the chapter on European Economic Community policies.

PROSPECTING AND EXPLORATION

The principal laws governing mining activity in Greece are the Mining Code of 1973 as amended by the law of 1976 and the Law on Mines and Quarries of 1984.

Geological services

The state has the right to carry out geological exploration and prospecting work throughout the territory. The aim of such work may be:

a) geological reconnaissance of the territory by aerial photography methods;

b) exploration of metalliferous regions using geophysical, geological and geochemical methods;

c) the realisation of surface work (drilling) in regions with a mining potential;

d) study and development of enrichment and metallurgical processing of ores in general.

The zones to be the object of basic geological survey are designated by the Ministry of Industry and published in the Official Gazette. The survey follows a predefined programme specifying the minerals to be sought, the duration of the survey and the methods envisaged, the amount to be spent and the company to undertake the work. This expenditure is financed by the state, through the public investment funds of the Ministry of Industry.

The Institute of Geology and Mining Exploration (IGME) is the organisation responsible for carrying out basic geological exploration and detailed geological surveys and for producing geological maps. The IGME is also responsible for research programmes in the field of geology and for managing the laboratories concerned.

Today exclusive mineral exploration and mining rights for the government are limited to hydrocarbons, solid fuels (coal, lignite), hydro-electric resources and radioactive minerals. A plan is being studied, however, concerning the extension of state control to other, perhaps all, minerals.

Lands availability

In accordance with a traditional arrangement in Greek constitutions, there are special laws governing the ownership and transfer of mines. Authorisation for mining exploration is a procedure that concerns only the applicant and the authorities (Ministry of Industry); third parties do not intervene despite any specific rights they may have.

In Greece, the holder of an exploration permit has a right of "quasi possessio, possessio juris" vis-à-vis the owner of the land and has a right to occupy the land for the purposes of mining exploration in return for a consideration. Restrictions on mineral exploration and mining are applied over selected areas such as national parks, archaeological and touristic sites and sensitive areas for national security. Restrictions on mining activities are considered on a case by case basis.

To the extent that the mine is declared to be "of public utility", several provisions of the Mining Code grant to the holder of mining rights priority vis-à-vis other rights existing on the site. In fact, according to the Mining Code of 1973 as amended by the law of 1976, the ownership of the mine is a property right separate from the ownership of the land.

Granting of exploration rights

A mining concession has to be promulgated by presidential decree. All contracts concerning the transfer of mining rights or the cessation of use or of usufruct become effective only after authorisation by the Ministry of Industry and Energy. The State holds the mining rights on metals wherever it carries out exploration work. This is a right of a general nature that can be exercised whenever it is deemed useful.

Exploration permits are granted by the regional administration (Prefect). Exploration permits are valid for a three year period which can be extended for one year. Permits are not granted to foreign nationals and entities outside of the EEC except if they receive an authorisation from the Ministerial Council.

Taxation

Mining exploration costs and related expenditures can be deducted either in the year during which they are incurred or spread over the following four years (Law of 1959).

Direct participation

Exploration carried out by the government may proceed regardless of the wishes of the owner of the land. It is done by the Institute for Geological Research, or, exceptionally, by third parties. The regions in which exploration is carried out by the state are designated "State Mining Regions". If prior mining rights existed, they are preserved until the rights cease for any reason, the region being handed over to the state as an integral part of its own mining region. The state carries out exploration in promising mineral areas and in public lands.

MINING AND METALS PRODUCTION

Mining

For the development of a mining project, the government requires from the mining company: (i) a detailed technical study for the Ministry of Industry; (ii) a study for the restoration of the site for the Ministry of the Environment, the Ministry of Industry, and the local authority; (iii) an impact study for the Ministry of the Environment, the Ministry of Industry and the authority responsible for forests.

The procedures for developing a mining project in Greece have been judged, in the report "Greek Minerals: The industry and its competitiveness" produced for Eurométaux in 1989 as being complicated, lengthy and costly.

Regulations

The granting of a mining concession is a procedure that concerns only the applicant and the Ministry of Industry, without third parties, regardless of any specific rights they may hold.

A mining concession is for a period limited to 50 years, which may in exceptional cases be extended by 25 years. However, the ownership of the mine is not protected by the Constitution. In the case of a state mine, the terms of the concession include compulsory renegotiation. Once a company has obtained a mining concession it has to proceed with work on the site or risk losing the concession. It has the right to exploit all the minerals on the site, except in the case of a reservation by the state. Once in operation, private producers of concentrates are obliged to give priority to supplying state processing plants.

In order to promote geological exploration by private mines, the Ministry of Industry may request the owner to present a work programme within three months. If this programme is considered unsatisfactory, it has to be replaced by one proposed by the Ministry. Where the owner disagrees, the government has the power to carry out exploration work without his consent. If the work leads to the discovery of exploitable deposits, the owner of the mine has to start mining within one year, reimburse the state for the exploration costs and pay an annual rent. Non-compliance with these procedures leads to the forfeiture of the owner's rights and transfer of the mine to the state.

Taxation

There are no special taxation rules for mining activities. The general 40 per cent tax rate applies to all firms including mining. This rate is reduced to 35 per cent for public firms quoted in the stock exchange and for those which have made productive investment after 1 January 1987. Exploration expenditures can be deducted from income for tax purposes during the year of realisation or in the following four years.

Enterprises effecting new productive investment have the right to make certain deductions from net pre-tax profits. In the case where the investment benefits from subsidies on the interest paid, the regime of tax deductions is not applicable.

Metals production

Inputs

Industrial consumers in Greece pay the highest electricity tariffs in the EEC. Furthermore, the shortage of electricity has very seriously limited the expansion of the aluminium and magnesite industries in Greece. Sometimes the energy producer and supplier, the Public Power Corporation, is not even able to assure supply to new projects. Aluminium de Grèce and Larco are the only enterprises to benefit from preferential tariffs (about 4.5 drachmae per kWh as compared with an average tariff of between 9 and 10.5-11 drachmae per kWh respectively for big and medium consumers).

Direct and indirect support

Due to high indebtedness of some Greek firms, the government decided to convert these debts into equity participation. This financial restructuring programme has affected the following firms: Larco, Fimisco, Macedonian Magnesites, the Eleusis Bauxite Mines, ELSI, MEDIOR and PORCEL. The Industrial Development Corporation was instrumental in this debt/equity conversion. The government policy is now to pursue a privatisation programme for state-owned firms. In this connection, the Law 2000/91 was voted which provides for the privatisation of state-owned firms, the simplification of liquidation processes and the reinforcement of competition rules.

The government can also intervene through interest rate relief on sums borrowed by enterprises for new productive investment. This amounts to a percentage equal to that of the aid authorised for the investment concerned. This subsidy is granted for the first three years. It is not granted if the investment is imputed to tax deductions. This rule does not apply to investments made by public companies and entities, but does apply to their subsidiaries and to companies set up by local administrations.

Direct participation

The Mining Code of 1963 did not allow the governments to acquire shares of mining companies. However, the Law 1262/1982 on "Incentive measures for the economic and regional development of the country" (Government Gazette vol. A/70 of 16 June 1982) provides for such equity participation under certain conditions. The government participates in the industry in several ways: directly, through companies such as HIMIC; indirectly, through the nationalised banks that hold shares in independent companies; and through the management of firms in difficulties.

The principal investments are:

a) HIMIC (Hellenic Industrial & Mining Investment Co.) founded in 1975 by four Greek banks, with the main objective of promoting the development of mining resources and renewable resources and establishing industries of national interest. At the end of 1985, the government decided to make HIMIC its main co-ordination agency for mining and metallurgical activities. It therefore took a majority holding in the capital of this company (37.4 per cent) together with four banks: National Bank of Greece (NBG), (23.5 per cent); Hellenic Industrial Development Bank (HIDB), (23.5 per cent); National Mortgage Bank of Greece (NMBG), (12.5 per cent) and the National Investment Bank for Industrial Development (NIBID), (3.1 per cent). HIMIC controls a substantial part of the mining and metallurgical activities in Greece, included Emmel, the lead producer; Hellenic Ferro-Alloys, a producer of chromite and ferro-chrome; and Metba, involved in geological exploration for lead, zinc and gold.

b) HIDB (Hellenic Industrial Development Bank) which controls the GEMEE (Project Studies & Mining Development Group) is essentially a mining exploration subsidiary. It has majority holdings in several enterprises, such as Hellenic Alumina (ELVA), a Greek-Soviet alumina production project and minority holdings in HIMIC and Metba.

c) NDBID (National Development Bank for Industrial Development), an association between the National Bank of Greece for 40 per cent and a consortium of international banks. NDBID has minority holdings in a number of mining and metallurgical companies such as Aluminium of Greece and Hellenic Ferro-Alloys. The NDBID has also been active in mining exploration through its subsidiary Mevior.

d) The Organisation for Problematic Firms has taken over the management of several private companies including Fimisco, Larco, MTM and others in 1983 due to financial difficulties.

HORIZONTAL POLICIES

Foreign investment controls

A foreign person or company with headquarters in Greece may be classified as foreign if the Council of Ministers determines that it is under the legal or financial influence of foreigners, taking into account the nationality of the majority of shareholders, the capital and other important factors. To the extent that foreign entities do not have the right to an exploration permit mining rights are linked to mining exploration rights), they cannot obtain a mining concession, except in an indirect form.

In order for foreigners to be able to obtain mining rights, prior authorisation by the Council of Ministers is necessary to modify the agreement. This may take different forms:

a) a contract transferring the mining rights or their usufruct;

b) the acquisition of shares in Greek companies that hold mining rights;

c) acquisition by succession or testamentary donation.

Authorisation may be refused or be made subject to conditions, non-compliance with which may cause forfeiture. Once operational, there is no discrimination between Greek and foreign companies: the two groups operate under the same conditions as regards access to mining products and are subject to the same taxes and may benefit from the same direct or indirect aid. Foreign investment in Greece is not very common. The most significant example in the non-ferrous metals sector is that of Péchiney in Aluminium de Grèce.

Regional development

Regional development assistance takes the form of subsidies granted under certain conditions, for new investment in excess of 400 million drachmae. Assistance may be as grants or equity participation.

The state participation which is taken only at the request of the enterprises. For investments of 400-600 million drachmae, a 50 per cent subsidy is granted without any conditions, and the other 50 per cent takes the form of equity participation in the share capital of the enterprise. For investments over 600 million drachmae, the state takes an equity participation in the share capital of the enterprise equivalent to 50 per cent of the investment.

Amounts granted depend on the region where the enterprise carries out its work. There is no subsidy for enterprises working in the central regions.

Research and development

The Ministry of Industry, Energy and Technology, is financing non-ferrous metals research through the Reinforcement Programme for Researching Potential and the Programme for the Industrial Research Development (PAVE). Firms which are pursuing applied research can create a tax deductible reserve fund up to certain limits.

Trade policy

As a Member of the EEC, Greece applies the common external customs tariff, the section of which relating to the six non-ferrous metals is summarised in the chapter dealing with the European Economic Community.

CONSUMPTION/SUPPLY

Recycling

Recycling of aluminium is supported under the Co-ordinated Programme for the Development of Wastes which is financed by the Ministry of Industry, Energy and Technology. Under the PAVE programme, studies were financed on the recycling of lead accumulators, other metals and the recovery of metals contained in tailings.

BIBLIOGRAPHY

Greek Mining Enterprises Association (1989), *Greek Minerals: The industry and its competitiveness.*

Grecofile (1988), "Greek Metal Industries, from Mining to Manufacturing", *The Greek Exporter*, No. 3.

US Bureau of Mines (1988), "The Mineral Industry of Greece", *Minerals Yearbook, 1987.*

Bundesanstalt für Geowissenschaften und Rohstoffe (1984), *Griechenland*, Hannover.

IRELAND

INTRODUCTION

Ireland is a significant producer and exporter of alumina and of lead and zinc concentrates. Alumina output through the Aughinish alumina plant (in full operation since 1987) comprises about 50 per cent of the value of the metals industry and production of zinc concentrates represents another 15 per cent.

Zinc production may be further boosted after the announcement in 1990 by Conroy Petroleum and Natural Resources that the feasibility study for its Galmoy project showed reserves that could support a zinc mine for at least ten years with a production of 70 000 tpy. The company is now applying for planning permission. The project might be developed in association with Outokumpu (20 per cent), which owns and operates Tara Mines, the largest zinc mine in Western Europe, located in the Midlands of Ireland.

The Government had traditionally exerted efforts to attract foreign capital to explore and develop mineral deposits in Ireland, as few local sources of risk capital have been available in the past. In the early 1970's, the Irish Government revised the orientations of its mineral policy, towards greater involvement in mining projects. State intervention established by the Minerals Development Act, 1940, was checked by Court decisions in the mid-seventies and since has been severely curtailed by the Minerals Development Act, 1979.

Ireland is looking to increase its role as a supplier of non-ferrous metals within the European Community. As a consequence, the country has been an active player in the recent discussions leading to the working out of a European mining policy.

All mining laws are supervised by the Minister for Energy who has broad regulatory powers; the Department of Energy is responsible for the development of the country's mineral resources and the Minerals Division of this Department deals with the administrative aspects of all non-fuel minerals. The Geological Survey of Ireland (GSI) is an office of the Department of Energy. Its Mineral Resources Division, in association with the Minerals Division of the Department for Energy, administers prospecting licences and mining facilities and provides technical information and advice to the Minister.

PROSPECTING AND EXPLORATION

The main legislative provisions concerning the exploration and development of non-fuel minerals are contained in three Acts, collectively termed the Minerals

Development Acts, 1940 to 1979. These cover all minerals with the exception of stone, sand, gravel and clay. They apply throughout the entire country, territorial waters and in the Irish designated continental shelf. The Minerals Development Act, 1940 (in substitution of the Mines and Minerals Act, 1931) comprises the most comprehensive part of this legislation and contains the main provisions in relation to the definition, the rights to prospect, the granting of prospecting licences, ownership of minerals, and requirements with regard to fees. The Minerals Development Act, 1940 also established the Mining Board, with appointments by the Minister. The Petroleum and Other Minerals Development Act, 1960, makes minor amendments to certain provisions of the 1940 Act, and the Minerals Development Act, 1979, has, as its main provision, the effect of vesting the exclusive right of working minerals, with certain exceptions, in the Minister for Energy. This Act also specifies the Mining Board's areas of competence. The safety and environmental aspects of mining and prospecting activity are also subject to the Mines and Quarries Act 1965, the Safety, Health and Welfare at Work Act 1989 and Local Authority Planning Regulations and Acts.

Geological services

The Geological Survey of Ireland (GSI) plays a pivotal role in the mining industry. It makes technical assessments for the Department of Energy in the process of granting prospecting licences and provides a broad range of information and services in areas such as geology, geophysical, geochemical and hydrogeological surveys.

The Databank and Industrial Minerals Section of the Mineral Resources Division in the GSI is active in promoting the minerals potential of the country. The Section has two functions: to collect and store data on mineral resources and to promote the geological study of the country to further aid the development of the mining industry. The Section is also involved in monitoring the development of resources. Private firms are an important source of geological information which is normally turned over to the Geological Survey.

Lands availability

Exploration and mining activities are prohibited in natural parks and designated natural monument sites.

Granting of prospecting licenses and exploration permits

Applications for State prospecting licences are submitted to the Secretary, Department of Energy, and then further transmitted to the Geological Survey.

A landowner may conduct preliminary prospecting on his own land (Minerals Development Act, 1979). For other parties undertaking such preliminary exploration, no special permission is required from the Minister although it is necessary to make arrangements with the landowners to work on their lands. However, if companies want to do more detailed geological exploration, they must apply for a prospecting licence and submit a report to the Geological Survey on the results of the preliminary survey carried out. Information therein is treated as confidential. A company may also establish its interest over an area with the Geological Survey by a written submission of a "statement of interest". Such a statement normally relates to larger areas for which the company expects to request a formal prospecting licence. In the same way, the results of the reconnaissance work made under a "statement of interest" are included with any subsequent prospecting licence application.

Exploration, other than preliminary or regional surveys, can only be undertaken by the holders of prospecting licences, who may be individuals, firms or joint ventures. Such holders have the right to undertake exploration in the licence area without the need to negotiate with private mineral owners, but should respect access to their lands. If the mining rights on a property have been well established as belonging to a private landowner, a company wanting to develop an exploration activity on this land must negotiate directly with the owner. However, often there is uncertainty over the ultimate ownership of the mining rights in the case of private lands. In this case, the company must ask the Government for an exploration permit that can be granted without a guarantee that the mining rights effectively belong to the state. If doubt persists, an arrangement between the mining firm and the surface occupier would be necessary.

The Minister for Energy has power to grant prospecting licences in respect of minerals which are not being worked or which are not being worked efficiently, whether the minerals are in state or private ownership.

The applicant must be able to demonstrate the financial and technical capability to carry out an acceptable prospecting programme on the allocated licences. In addition, an insurance bond, currently Ir£ 1 000, must be provided for each licence from which compensation for any damage caused by the prospecting activity can be paid. The initial term of a licence is for two years during which the applicant must carry out a work programme involving a minimum expenditure of Ir£ 5 000. The prospecting programme must be carried out in accordance with the standard terms of the licence. If the applicant performs satisfactorily he may apply for renewal of the licence. At the end of the initial period and each renewal period, full reports must be submitted to the Department of Energy on the progress of the prospecting programme. In general, the work programme expenditure requirements increase with each renewal with significant expenditures being required from the end of the sixth year onwards. A small fee is charged for a prospecting licence.

Tax incentives

Exploration expenses are considered capital expenses when related to the research and development of an orebody, construction of means of access, geological expertise expenditures, etc. The Finance Act, 1974 instituted a tax allowance of 20 per cent of expenses incurred in mineral exploration, but which was abolished by the 1990 Financial Act.

Under the Programme for Economic Expansion, funds have been allocated for Technical Assistance Grants towards the costs of prospecting for, and development of, minerals by private interests. Where commercial deposits of minerals are likely to be workable in an area and their development is desirable in the national interest, the Minister for Energy may consider applications for financial assistance in respect of suitable schemes of exploration or development. General conditions include:

i) the grant is payable in respect of each approved project, up to a limit of one-half of costs;

ii) while grants may, in suitable cases, be payable for development work as well as for exploration, expenditure on the acquisition of plant machinery, equipment and similar development items may not be included for the purpose of grants;

iii) in the event of commercial production arising out of a particular prospecting or development project, a special royalty related to the value of the minerals produced is imposed to cover, as far as practicable, the total amount of the grant made towards the costs of the project.

Provisions of the 1990 Finance Act allow mineral exploration expenditure to be totally offset against other taxable income of the same company. On the other hand, the government has yet to spell out fully the taxation rules concerning mineral development activities.

MINING AND METALS PRODUCTION

Mining

Applications for Mining Leases and Mining Permissions are submitted to the Secretary, Department of Energy.

Regulations

The Minerals Development Act, 1979 vests the exclusive right of working minerals, with certain exceptions, in the Minister for Energy. The Minister is empowered either to work the minerals, or to grant a lease or licence for their development. As a consequence, the Minister for Energy may grant authority for either State-owned minerals or for privately owned minerals in respect of minerals for which the mining rights are vested in the Minister through the Minerals Development Act, 1979. This is done through State Mining Leases, Permissions and Licences, the specific facility depending on the scale of the envisaged working and the mineral ownership. Terms are normally negotiated on an individual basis.

The 1979 Act does not interfere with acquired rights, and mainly concern those deposits which were being worked by the mineral owner on 15 December 1978. To grant a State Mining Licence, the Minister is required to give public notice of intention and to recognise the right of the mineral owner(s) to apply to the High Court to settle the question if they are aggrieved. An exclusive vested mining right acquired by the Minister may be surrendered at any time subject to approval by the Minister of Finance. In the cases where there is no agreement on the compensation to be paid to owner(s) of private minerals which the Minister proposes to develop, the Mining Board determines the amount of such compensation, always subject to the right of the mineral owners to apply to the High Court. Mining Board powers were specified by the Minerals Development Act, 1979.

The types of authority the Minister may grant are:

— State Mining Leases: These are granted in respect of state-owned minerals and are normally issued against an operational rent or royalty. Financial terms are determined by the Minister in relation to each lease application, evaluation of the deposit size, etc.

— State Mining Permissions: These are mining facilities of more limited and specific scope than leases. They are granted for state-owned minerals, but only permit the removal of a restricted quantity of the mineral for a limited period. Rent or royalty is charged on the same basis as for State Mining Leases.

— State Mining Licenses: These are granted by the Minister for privately owned minerals, the mining rights to which are vested in him through the Minerals Development Act, 1979.

In addition to the need for a state mining authority, potential developers are also required to obtain planning permission from the local planning authority and any necessary surface or access rights from the land owner(s). The application for planning permission includes details of waste disposal, provision for restoration and other environmental considerations for an environmental impact assessment. Each lease is the object of an "ad hoc" negotiation with local authorities and planning permissions are issued under the various Local Government Acts. Licences must also be obtained from the Local Authority for effluent discharge under the Water Pollution Acts, 1977-1990 and Air Pollution Act, 1987. Either the developer or third parties may appeal the decision of the local authority to an independent National Planning Board whose Chairman is a judge of the High Court. The local authority is required to notify the appropriate bodies if features of special interest (e.g. tourism, archeological site) are involved.

Currently, there is an application fee of Ir£ 5 000-10 000. Applicants are also required to meet the cost of any consultants engaged by the Minister in connection with the consideration of any aspect of the application. Rental and royalty payments are agreed on a case-by-case basis with each lease applicant, as are any other fiscal terms and fees such as environment monitoring expenses.

These provisions of the Minerals Development Act, 1979 are aimed at preventing difficult situations arising from the historical and general principle in Ireland bestowing the mining rights on the owner of the land. Indeed in Ireland, several land reforms since the end of the nineteenth century have created a complicated situation, because the mining rights were often effectively separated from the ownership of the land. The result is that today they may belong either to the private landowner or to the state even though normally the land over the minerals is privately owned. State ownership of minerals derives mainly from provisions of the Lands Acts 1903 to 1923 that cover the sale of lands through the Irish Land Commission and provide for the retention of the mineral rights by the State in certain circumstances.

Overall, it has been estimated that 65 per cent (by area) of all minerals are state-owned, including all gold and silver[1]. The remainder are privately owned. Normally, the land over the minerals is owned privately. The sole right to develop both state and privately owned minerals rests with the Minister for Energy and, through him, the holder of the relevant mining lease. Development normally takes place with the consent of the private mineral owner or landowner. Payments of compensation by a developer to mineral owners is made on a fair and equitable basis.

The Minerals Development Act, 1979 replaced the procedure of the Minerals Acquisition Order, regulated by the 1940 Law, which permitted the state acquisition of the mining rights in cases of doubt about ownership. Before the government had the right to compulsory acquisition orders in cases of uncertainty about the ownership of the mining rights. Failure to use this procedure in the case of Tara Mines had major effects on the evolution of Ireland's mining law leading to a most important court case which lay at the origin of the abolition of the Minerals Acquisition Order. Government intervention in the mining industry was the theme of a long debate in the Irish press and in Parliament during 1976 and 1977 and is the principal concern behind the Minerals Development Act, 1979.

Tara Mines had explored an area on the basis of a general licence that proved to be a rich orebody. The company asked the government for a mining lease for the entire zone, knowing that some parts of it belonged to a private owner. In the meantime, Tara Mines tried to purchase the private land but another company, Bula, offered a higher price and became the owner. Tara notified the Government of the new situation that reduced in practice a part of its mining rights that had been accorded to the whole Navan mine. The Government then issued a compulsory acquisition order on Bula to recover the mining rights that had been leased to Tara. Bula went to Court arguing that the compulsory acquisition order was illegal and won its case first in the High Court and later in the Supreme Court. As a consequence, Tara Mines received the mining rights over only a part of the Navan mine that it had originally discovered; subsequently it was forced to accept a state participation of 25 per cent in the mine. The Government participation has since been sold to Outokumpu the principal owner of Tara Mines in 1989. For its part, Bula took as partner the private landowner and had to sign an agreement with the Irish Government, according to which the state received a 45 per cent share, of which 25 per cent was without compensation and 20 per cent bought at a negotiated price. Outokumpu has since bought out Bula's mining rights.

Minerals taxation

The general corporation tax rate is 50 per cent[2]. There is no special tax rate for mining operations. No local or provincial taxes are imposed. The compensation for the lease of a mining right belonging to a private owner is negotiated directly with the mining company and is often proportional to the commercial value of the mineral exploited. When the mining rights belong to the state, the lease sets out the conditions of remuneration for the state whether under the form of a royalty or a percent of net profits.

There are no withholding taxes on dividends on the remittance of branch profits. Investments from the United Kingdom have been subject to special tax treatment. Ireland has income tax treaties with several countries that reduce or eliminate the normal Irish withholding tax for payments on interest, patents and mineral royalties.

Tax incentives

Any capital expenditure incurred in mine development may be used to qualify for a tax allowance. The amount of the tax allowance was previously determined by reference to the amount of capital expenditure, the estimated life of the mine and the residual value of the assets; it could not exceed 20 years. The costs of acquiring mineral leases were depreciable on a straight-line basis over the life of the mine or 20 years, whichever is less, beginning with the first year of production. Traditionally, up to 50 per cent of exploration and development expenditures within Ireland have been deductible.

The 1990 Finance Act made significant changes in taxation schemes: it allows mineral exploration expenditure to be offset against other taxable income of the same company and losses can be carried forward. Otherwise there are no special incentives. Mining is subject to the full company taxation rate and there are no longer provisions for accelerated depreciation. The total cost of mining plant and equipment may be depreciated for tax purposes in the first year in which used. Road vehicles are depreciable at 20 per cent on a reducing balance method, whereas depreciation of non-commercial vehicles is restricted. The cost of mine buildings may be deducted as incurred. Housing, medical and educational facilities are generally not depreciable if their useful lives are longer than the life of the mine. Net operating losses may be carried back one year and carried forward indefinitely. The sale of a capital asset may be taxable as a capital gain.

Metals production

Direct and indirect support

The Government's Programme for Economic Expansion in 1958 provided a major impetus in the industrialisation of the country by offering a package of incentives to domestic and foreign companies. Irish Government policy to encourage private activities led to the establishment of two agencies: the Industrial Development Authority (IDA) and the Shannon Free Airport Development Co. Ltd. (SFADCO). IDA was established by the Government with the brief to create wealth and employment. It aims to attract domestic and overseas investment offering financial incentives for capital and personnel. IDA is fully financed by the government.

A major incentive offered by IDA is a corporation tax of 10 per cent for manufacturing and financial services, guaranteed to the year 2010. Other financial benefits include depreciation allowances enabling buildings and machinery to be written off against tax and the repatriation of profits without hindrance.

The Shannon region comprises the counties and areas of Clare, Kerry North, Limerick, North Tipperary and west and south west Offaly. It is the only region in Ireland with a dedicated government agency, SFADCO.

The concept of a duty free zone was pioneered in the area following the establishment of the Shannon Free Zone in 1959, the precursor of all duty free shopping developments. The Shannon port is now a duty free port and SFADCO is trying to attract processing operations for metallic products to the area through duty free access for goods in process, no duties levelled on goods destined to non-EC countries, and duty free into bonded warehouses.

An example of a minerals processing company which has located in the area is Aughinish Alumina. The plant is the largest chemical concern in the country at a development cost of US$ 1 000 million and represents the biggest single private investment in the country.

With the aim to co-operate with private interests in promoting industrial mineral-based industries, the Industrial Minerals Development Group has been established, a loose affiliation of various interests including the Industrial Development Authority, the Ministry for Energy and the GSI.

Government ownership

The most important participation of the Irish government in operating mines was its 25 per cent share in Tara Mines that was sold to Outokumpu in 1989. All gold and silver mines are state-owned.

HORIZONTAL POLICIES

Foreign investment controls

The Irish Government may require equity participation in mining operations. Historically, it seems that no specific controls on foreign investment have been in place, even if varying phases in the country's general openness to host foreign investment can be distinguished. Currently all licence applicants are assessed on the basis of technical capability and financial capacity to complete work commitments.

The period from the fifties to the earlier seventies may be characterised by a relative openness to foreign firms wishing to invest in order to develop the domestic mineral resources. During the seventies the country went through a protectionist period and the government exercised its option to take an equity share in Ireland's most important mine, developed at that time by a foreign group. The sale in 1989 of the government's participation in Tara Mines to Outokumpu brought to an end the state's direct intervention in operational mines.

Environment

Up until 1990, mining activities in Ireland were subject to the Local Government Planning Act, 1963 and 1976 and to the Water Pollution Act of 1977. A Planning Permission is always required for any operation or development having an environmental impact on the site of a mine. In Ireland this is a major issue because it is possible for a third party to challenge the decision of the Planing Board through a High Court Action and even through the Supreme and European Courts.

Reflecting the EEC Directive 337/85 on environmental impact assessment, the Irish Planning and Development Regulations, 1989-90 are the major legal framework in which to deal with environmental issues for mining. The EIA process requires prospectors to compile a full environmental audit allowing the public and other interests to comment upon it, before planning permissions are given.

The new regulations establish a formal procedure for applying an EIA to the proposed development by or on behalf of the local authorities. The procedure is essential to the planning process of any mining development and it involves three stages. First, an environmental impact statement is prepared by, or on behalf of, the local authorities, normally by a neutral party; secondly, the Ministry for Energy considers the report taking into account any submissions or observations that have been made by other interested parties; thirdly, if permission is granted the operators must guarantee that all development can be carried out as laid down in the EIA. The environmental studies are designed to ensure that any proposed mining activities will meet more rigorous standards. These arrangements are complementary to the normal existing procedures and requirements of the local planning authorities.

Mining companies' choice of environmental expertise is, via the Local Planning Authority, ultimately subject to the approval of the Department of Energy. This Department also uses the Geological Survey of Ireland to assess the adequacy of any environmental statements submitted to it.

An Environmental Baseline Study (EBS) covers all aspects of the current environment including rivers, vegetation, air quality, socio-economic structure, human and animal populations and agriculture. When the EBS is complete, the mining and processing methods are superimposed on the current environment and all possible impacts are assessed, i.e. an environmental impact assessment is done. A third section includes controls for water quality, dust and vapour emissions, noise and effect on other industries in the area such as tourism, marine farming and agriculture.

The Irish government is in the process of setting up an Environmental Protection Agency to work independently of the government, acting as consultant to determine a project's environmental viability. The Environmental Protection Agency will have the following functions:

i) control and regulation of development likely to pose a major risk to environmental quality;

ii) general promoting of environmental quality;

iii) support, back up and advisory services and;

iv) environmental research.

The Agency will also have direct responsibility for determining licences, monitoring emissions and discharges and any enforcement action associated with new development that has a potential for pollution.

Following the introduction of new regulations, any prospective mining operation is to take account of the following key issues/impacts:

i) impact on aquatic systems;

ii) visual/landscape impact;

iii) atmospheric emissions;

iv) noise/vibration;

v) agricultural impact;

vi) rehabilitation/decommissioning (site restoration plans are also required by the Planning Authority);

vii) socio-economic impact.

Research and Development

The Irish Science and Technology Agency (EOLAS) is a state-sponsored organisation whose objective is to promote the application of science and technology in industry. The Science Division provides specific technical terms in minerals development, minerals processing and the industrial application of minerals.

Trade Policy

As a Member of the EEC, Ireland applies the common external customs tariff, the section of which relating to the six non-ferrous metals is summarised in the chapter dealing with the European Economic Community. Licences are required for the exportation outside of the EEC States of waste and scrap of aluminium, copper and copper alloys, lead, and ash residues of copper and copper alloys. Exportation to places other than Northern Ireland and Great Britain of certain non-ferrous ores and metals (including alumina) also requires licences.

CONSUMPTION/SUPPLY

Exploration abroad

The general drop in mining exploration in the 1980s led to the withdrawal of a number of international companies from Ireland, which in turn resulted in reduced availability of funding and decreased numbers of experienced geologists.

Since 1983, major efforts have been undertaken by the domestic exploration community and the Department of Energy to promote mineral exploration by private enterprises in the country in a spirit of close knit collaboration. As a consequence, and although with minimal government financial support, a number of junior Irish companies have been formed that successfully raised over IR£ 50 million to date from national sources, as well as several successful consultancy organisations. There have been some successful ventures by Irish companies involved overseas, now at an advanced stage of assessment.

The two principal companies with interests abroad are Kenmare Resources PLC and Crowe Schaffalitzky & Associates Ltd. (CAS), both basically involved as geological consultants. The first is developing its activities in Mozambique (mineral sands), Philippines (gold) and Sudan (auriferous tailings); the second's activities are concentrated in Australia, Turkey, and Gabon.

Stocks

There is no government-run stockpile of non-ferrous metals.

Recycling

Ireland is committed to raising the recycling levels of all waste, including non-ferrous metals, to the optimum level. The Department of Environment operates a scheme of 50 per cent capital grants to encourage the development of local recycling projects. It is in continuing liaison with representatives of the beverage can industry to establish a national structure for beverage can recycling operations. The success of these initiatives can be gauged by the can recycling level in Ireland which has risen from a position of virtually nil in 1986 to 10 per cent in 1991 and which continues to rise.

NOTES AND REFERENCES

1. NESC Report, Minerals Policy, Dublin, 1981.
2. Until 1974 the Acts exempted mining operations from income tax and corporation profits tax for a period of four years from the date on which production commenced; for each of the next subsequent four years 50 per cent of the profits were also exempted. This concession applied to copper, zinc, pyrites, gold and silver deposits exploited by any new enterprise commencing production within a period of ten years from April, 1956. The abolishment of these tax incentive schemes followed criticism of government policy and the emergence of a nationalist wave.

BIBLIOGRAPHY

HORNE, R.R., GARDINER, P.R.R. and PYNE, J.F. (1989): "Mineral Exploration and Development in Ireland: Its Administration and Infrastructure", *Geological Survey of Ireland*, Information Circular 89/1.

International Mining, June 1990.

Mining Journal, 24 August 1990, Special Supplement on Environment.

ITALY

INTRODUCTION

The mining of non-ferrous ores in Italy is now restricted to lead and zinc. While metal output in Italy is high, particularly where zinc, lead and copper are concerned, it does not cover national consumption in the case of any of the six main non-ferrous metals.

A specific feature of Italian Government policy on mining and metallurgy, as it applies to the non-ferrous sector, is the crucial role played by state holdings in the growth and restructuring of industrial activities. Indeed, because of the difficulties encountered by Italian firms in the sector due to the two oil crises, the state eventually took control of almost all the major companies, grouping them into holdings, which enabled them, albeit slowly and gradually, to restructure their industrial operations. There are three major state holdings in the non-ferrous metals sector: SAMIM (Società per Azioni Minero-Metallurgiche) for lead and zinc; Sameton, for the recycling of non-ferrous materials; and Aluminio Italia (now known under the name of Alumix) for the aluminium product group.

Italian legislation governing mines was first consolidated by Decree No. 1433, of July 1927, which replaced the laws of the old states predating the establishment of the Kingdom of Italy (1861). With regard to mines, quarries and peateries, the 1927 statute was supplemented by another, numbered 1360 of November 1941, on the classification of mineral substances and by Presidential Decree No. 620, of June 1955, providing for the administrative decentralisation of the departments of the Ministry for Industry and Trade. On the territory of the Autonomous Regions with Special Status — Sicily, Sardinia, Trentino Alto-Adige and Valle d'Aosta — this question is governed by different regional statutes.

Pursuant to the legislative provisions indicated above, only the state (or Autonomous Region) has the authority to dispose of mineral resources and to allocate them to persons offering the best assurances and having the technical and financial capacity to carry out exploration work and to exploit deposits.

The minerals classification system governs exploration for, and the extraction of, mineral resources. It divides them into two categories: mines and quarries. Exploration for and the extraction of metal ores, metalloids, solid fuels, mineral waters, vapours, gases etc. fall within the category of mines. Quarries remain under the sole jurisdiction of the local authorities.

Within that category, however, a secondary classification distinguishes between minerals of national interest and minerals of local interest. The administrative provisions relating to minerals of national interest fall within the jurisdiction of the Directorate General for Mines of the Ministry for Industry and Trade. Non-ferrous metals belong to this group. The provisions governing minerals of local interest, with some limitations, fall within the jurisdiction of the Mineral Products Bureaux for the districts in question.

The interventions introduced in the mining sector by the Law of 1988 were followed up by Law No. 221 of 1990. In addition, the new Law brings a certain number of significant innovations with respect to the preceding legislation, among which the adaptation of certain Community interventions, the extension to private enterprises of support for mining activities abroad and the introduction of the obligation for the restoration of the environment at the operators' expense.

The fundamental importance of mining policy for Italy is to be seen in its function in ensuring the limited and reasonable supply of mineral raw materials required by the processing industries downstream of the extractive sector. Intervention also appears essential for the continuance of a mining culture, seen both as the main work activity in certain disadvantaged areas of the country and as know-how of the national industry in terms of capabilities for programming and operating on the international market.

PROSPECTING AND EXPLORATION

Geological services

The Directorate General for Mines of the Ministry for Industry and Trade comprises administrative and technical services, a statistical service, the national office for hydrocarbons, the geological service and the chemicals service. It has responsibility for the application of state mining laws and controls and monitors mining operations. The Servizio Geologico d'Italia (Italian Geological Service) was subsequently transferred to the Ministry for the Environment[1].

The main task of the administrative services is to issue mining concessions, while the special role of the technical services is to exercise technical supervision. The statistical service processes the data received by the bureaux in the field. The geological service is responsible for producing the National Geological Map, while the General Directorate of Mines in the Ministry of Industry intervenes for basic surveys on the national territory and operational surveys. Active collaboration with university institutes has been established.

In accordance with the legal organisation of the autonomous regions, the legal and administrative authorities of these regions have been made independent of the central government administrations.

Lands availability

The only restrictions on the pursuit of mining activity arise from the laws setting up nature reserves and national parks which cover 4.3 per cent of the territory. Under Italian law, the owner of the land referred to in an exploration permit cannot oppose prospecting work.

Concessions and exploration permits

Any person holding a permit issued by the Ministry for Industry and Trade, in the case of minerals of national interest, and by the Chief Engineer of the Mining District, in the case of minerals of local interest, is entitled to prospect for mineral resources.

An exploration permit is issued to any physical or legal person requesting it who can demonstrate to the full satisfaction of the Administration that he has the necessary technical and financial capacity. The permit is issued for clearly specified time period and is restricted to particular locations and minerals. The maximum duration of the permit is three years, though on expiry it can be renewed. Once the permit has been obtained, the company must pay the state a duty in proportion to the number of hectares covered by the activity, and the permit is not transferable without prior authorisation from the Ministry. It is also obliged to pay compensation for damage done by exploration within the area covered by the permit. The owner of the land on which exploration is carried out may demand a guarantee.

Taxation

Under legislation which came into force in 1982, mining enterprises may qualify for government credits and subsidies ranging up to 70 per cent of the expenditure incurred in the course of operational exploration. In the event of a commercially viable find, the subsidies received must be repaid from the third year of commercial exploitation of the deposit. However, as regards tax relief, the mining industry is no different from other types of industry and there is no tax incentive for the sector.

Direct participation

Law No. 752/82, amended and completed by Laws 246/84, 399/87, 41/89 and 221/90, provides that basic research be carried out directly by the Ministry for Industry, Trade and Crafts, or by the ENI (Ente Nazionale Idocarburi) or, for activities relating to the territory of special status regions, by mining entities and enterprises of a regional nature whose competence has been recognised, separately or in co-partnership with the ENI, on the basis of specific agreements.

For the activity of operational exploration in Italy, it is planned to grant a subsidy of up to 70 per cent to holders of exploration permits or mining concessions for the land on which the exploration programme takes place.

For the activity of mining exploration abroad, it is planned to grant to the ENI and the Institute for Industrial Re-organisation (IRI), for minerals of steelmaking interest, and to mining entities and enterprises of a regional nature whose competence has been recognised, and to companies holding mining concessions on the national territory, possessing the technical and economic qualifications necessary, subsidies of up to 70 per cent of the costs incurred abroad for operational exploration work and shares in mining exploration activities already commenced.

Finally, special mention should be made of the Italian Government's participation since 1980, through its mining subsidiaries, in the exploration of sea-bed nodules on the Italian continental shelf by way of an interest in the Ocean Mining Associates (OMA) consortium.

MINING AND METALS PRODUCTION

Mining

Regulations

In the document granting a mining concession, the state may reserve the right to take a share of the profits in the company and to attach special conditions to the concession. Moreover, the structures necessary for the exploitation of the deposit and to ensure safety inside the perimeter of the concession are declared to be of public utility. When such structures have to be erected outside the perimeter of the concession or are necessary for its exploitation, the declaration of public utility is issued by the Ministry for Industry and Trade which, in cases of emergency, may order temporary occupation, specify provisional compensation and direct the payment of a deposit.

Accident prevention in mining operations is covered by Presidential Decree No. 128 of April 1959, providing for the policing of mines and quarries.

Taxation

The holder of a mining concession must pay the state a duty in proportion to the number of hectares contained within the perimeter of the concession. Apart from this duty related to area, an additional duty is payable the amount of which is not laid down by law but is fixed on a case-by-case basis. Taxation on the operating results of companies averages 46.4 per cent, comprising a central government tax of 30.2 pour cent and a local tax of 16.2 per cent on profit after initial assessment.

Metals production

Inputs

The subsidies currently available relate mainly to new investment projects in southern Italy, which qualify for reductions in rail freight rates and electricity prices. Where the other regions are concerned, rates tend to align with those of other industries.

Companies investing in the construction of new production capacity or the extension of existing plants in southern Italy qualify for subsidies ranging up to 40 per cent of the value of the industrial buildings, machines and equipment. The amount of the subsidies decreases with increasing volume of investment to a minimum of 15 per cent for investment over 15 billion lire. In severely economically depressed regions, subsidies may be greater but may not exceed 86 per cent of the investment in fixed assets.

Moreover, in addition to the reduction in rail freight rates and electricity prices, new investment projects in southern Italy qualify for reduced social security charges and for loans at subsidised interest rates (about 30 per cent below market rates) on 40 per cent of the investment in fixed assets up to a limit of 15 billion lire. Projects located in this region are also eligible for significant tax advantages.

Finally it should be pointed out that the holder of a mining concession may take out mortgages on the concession, subject to authorisation by the Ministry for Industry and Trade, as security for credit extended for the purpose of developing mining operations.

Indirect participation

The state enterprises have often passed through periods of great financial difficulty, but their activities have been safeguarded on economic and social grounds. The state has intervened in the Italian mining and metallurgical industry in such circumstances in order to ensure the preservation and financial restructuring of enterprises under its control in the sector, either by direct capital subsidies or by way of loans granted by state banks at subsidised interest rates.

Thus, for example, the state enterprise Aluminio Italia (now known as Alumix) on several occasions in the period 1984-85 received injections of public money in order to overcome operating losses and to implement a restructuring programme in the Italian aluminium industry.

During 1989, the Interministerial Committee for the Co-ordination of Industrial Policies (CIPI) decided on a programme for the zinc sector. It was proposed to create a national company structure in the zinc sector or in any event to seek effective co-ordination of the public presence.

Government ownership

Public and private enterprises exist side by side within the structure of the Italian mining and metallurgical industry. However, in view of the economic and financial difficulties experienced by Italian enterprises in the sector from the start of the 1970s, the Italian Government has been obliged to intervene in order to ensure the survival of the industry. Rather than achieve this through subsidies, the

Government opted to take the firms in difficulty into ownership by a large public undertaking.

After the dissolving of "l'Office autonome pour la gestion des industries minières et métallurgiques (EGAM)", state intervention in the mining and metallurgical sector is through companies of the ENI group: Nuova Samim operates in the metallurgical industry and SIM (Società Italiana Miniera) in mining.

At the start of the 1970s, the ENI group took over the share capital of 33 enterprises either directly owned by the state or managed directly by one of the major public undertakings controlled by the Ministry. This Group has been one of the chief instruments of Italian policy in the mining and metals sector.

EGAM reflects an Italian State tradition which had already brought about the creation of the Institute for Industrial Reorganisation (IRI) whose subsidiaries, Fincantieri, Finsider and Finmeccanica, for example, cover large parts of the Italian shipbuilding, iron and steel and metallurgical industries. The mining and metallurgical operations under state control, however, were transferred to SAMIM in 1978. In 1989, SAMIM was renamed Nuova Samim and operates in the metallurgical sector of the non-ferrous industry.

Moreover, Law No. 136 of February 1953 set up the "Ente Nazionale Idrocarburi" (ENI), a public law undertaking whose task is to adopt and implement initiatives in the national interest in the field of hydrocarbons through the intermediary of enterprises under its control.

The Italian State has not restricted its efforts to dealing with the management problems encountered by enterprises either controlled or wholly owned by it, but has endeavoured to fill in gaps and resolve structural difficulties with which certain sectors of Italian industry were constantly confronted in the same way as the mining and metallurgical industry. The new legislation adopted by Parliament in 1982 thus strengthens the scope for state intervention in industry. Legislative interventions in the extractive sector introduced by Law No. 752/82 were followed up by Laws 41/89 and 221/90.

In view of the financial difficulties encountered by EGAM and SAMIM, ENI recently had vested in it responsibility for all mining activity under state control in Italy and currently controls almost the whole of the Italian non-ferrous metal mining and metallurgical industry.

In 1986 the restructuring of this sector continued with the removal of mining and metallurgical activities from the subsidiary, SAMIM. The mining operations of this undertaking were transferred to Società Italiana Miniera, a subsidiary of Azienda Generali Italiana Petroli-Miniera SpA (AGIP Miniera). In the same way the metallurgical activities were transferred to a new company, the Nuova Società per Azioni Minero-Metallurgiche (Nuova Samim), which has also taken over the recycling activities previously in the hands of the subsidiary, Sameton.

In parallel, the interests held by the Italian State in the aluminium industry (alumina and aluminium plants) have been regrouped within the holding company, Alumix.

Taxation

Investment projects in the mining and metals industry qualify for accelerated depreciation for the first three years in the life of an asset up to a maximum of 15 per cent per year of the historical investment cost.

Also, investment in new projects launched in southern Italy qualify for a 50 per cent reduction in tax on income (assessed at 12.5 per cent) over a period of ten years from the date the company was set up. Such investment may also qualify for exemption from local income tax (15 per cent) for a period of ten years from the first year in which the company returns a profit. In addition, profits reinvested in the region may be exempted from local income tax up to 70 per cent of the total value. However, these concessions vary with the size of the enterprise, the location of the investment and the type of operation envisaged. In certain parts of the Trieste region, companies qualify for exemption from local taxation on income for a period of ten years.

HORIZONTAL POLICIES

Foreign investment controls

The new Law No. 221 provides for the possibility of granting exploration permits and mining concessions to individuals or legal entities of other states where there are reciprocal arrangements vis-à-vis Italian subjects. However, the required technical and economic capacities must be demonstrated and technical and administrative structures appropriate to the activities envisaged must be committed or available.

Environment

Apart from the area of hydrogeology, for which national directives exist, pollution control is regulated by each of the regional governments. The countryside protection laws do not refer specifically to mining. However, the laws setting up nature reserves and national parks do impose restrictions on mining activity.

There are no strict rules for the control of non-ferrous metal emissions to the atmosphere which, pursuant to Decree No. 322 of 1971 falls within the jurisdiction of the local authorities. The limits are set individually for each plant and vary from region to region. For dust emissions, the limits applicable to a plant located in Lombardy are about 10 mg/m^3 for dusts in general and 3.3 mg/m^3 for lead. The maximum quantity of lead permitted in the air of communities in the vicinity of metallurgical plants has been set under the same provisions at 10 mg/m^3 for an average of eight hours. The Ministry for the Environment instituted, by the Decree of 26th April 1989, the national register of special wastes, which also provides for a survey, using special methods, of wastes from the mining and metallurgical industries. The Ministry for the Environment, by the Decree of 12 July 1990, laid down standards for the limitation of polluting emissions from industrial plants and set maximum authorised values for emissions. Furthermore, Decree No. 214 of 28th May 1988 brought into effect EEC Directive 85/210/EEC relating to the lead content of petrol.

The limits for lead at places of work are controlled by the autonomous local authorities; maximum levels are 0.15 mg/m^3 air (for an average of eight hours). The limit for lead in the blood of workers is also monitored by the local authorities and fixed by the Italian Medical Association at 70 µg/100 ml.

Research and development

The subsidies available for applied research may be used either for projects to develop new technologies or for training (research) staff. The programmes concern the development of innovative and strategic technologies, likely to reach the industrial stage in the medium term (Decree of the Ministry for the University and Scientific and Technological Research of 29th May 1990).

For the non-ferrous metals sector, these innovations concern: the development of composite materials; vacuum depositing techniques; technologies for recovering metals from industrial waste.

In the mining and metallurgical sector it is planned (CIPI deliberation of 22 December 1983) to intervene in the following areas: reduction of exploration costs, increased productivity in the extractive industry and improvement of safety conditions; experimental studies on conventional enrichment systems in order to maximise yields and on new pyrometallurgical methods.

Trade policy

As a Member of the EEC, Italy applies the common external customs tariff, the section of which relating to the six main non-ferrous metals is summarised in the chapter dealing with the European Economic Community.

CONSUMPTION/SUPPLY

Securing overseas supply

Until the beginning of the 1980s, the Italian State did not really have a policy on overseas supply, since the Italian industry depended almost exclusively on a few suppliers. Since then, and in view of the total absence of Italian participation in the operation of foreign mines, the state enterprises have become the main focus of a supply policy.

SAMIM began to undertake exploration work abroad, mainly in co-operation with foreign partners. The target countries were Canada, Australia and Peru. The company very soon turned its efforts to seeking participations in existing companies in order, as soon as possible, to secure coverage of a proportion of the Group's concentrate requirements. Because of financial constraints this policy could not really bear fruit until after 1987, when the company's activities had been taken over by AGIP Miniera. The co-operative arrangements so far set up operate in Australia and relate to deposits under development, such as those at Thalanga (copper-zinc) and Mount Keith (nickel). In both projects the company is co-operating with the state-owned Finnish firm, Outokumpu.

The Mining Law of 1982 also laid the basis for a proper supply policy geared to co-operation with certain developing countries. Under these arrangements the Italian Government seeks to intensify action along lines of developing the mineral resources of certain countries in exchange for the establishment of supply commitments.

For interventions abroad, the new Law No. 221 on mining policy provides for the possibility of granting subsidies of up to 70 per cent of the cost of exploration activities and of taking shares in activities already under way. These subsidies may be granted, besides to the ENI, to mining enterprises of a regional nature and to private operators provided that they are holders of mining concessions in Italy.

Stockpiles

There is no systematic policy in Italy for the stockpiling of non-fuel mineral raw materials.

Recycling

The Italian Government is also developing schemes for the recovery and recycling of non-ferrous metals. In 1983, SAMIM's four recycling plants for aluminium, copper and lead/zinc were merged with the five sites of the Tonolli Group in this field to form a new company, Sameton. However, due to severe financial difficulties experienced by Sameton, the Tonolli Group withdrew from the company.

NOTE

1. Act No. 349, 8 July 1986 establishing the Ministry for the Environment and Standards for Related Environmental Damage, la Gazzetta Officiale (Official Journal) No. 56, 15/7/86; D.P.R. 15 January 1987 (G.U. No. 71, 26/3/87).

BIBLIOGRAPHY

Eurométaux (1989), *Conditions of Profitability of the EC Mining Industry*, Brussels, November.

Law of 6 October 1982, No. 752 and of 1990, No. 221.

SAMIM, *Annual Reports*, several years.

JAPAN

INTRODUCTION

In the initial post-war years, Japan's main objective was to rebuild its shattered domestic industrial base. Its domestic resources of many major non-ferrous raw materials were adequate for its needs, and smelting and refining centres were based on indigenous ores. It imported raw materials only for aluminium, nickel and tin. By the early 1960s, however, domestic demand for non-ferrous metals had in most cases out-stripped the production capabilities of domestic mines and import dependence had increased.

During that period, Japan successfully built up a modern and highly efficient world class non-fuel minerals processing sector that consisted mainly of primary aluminium smelting, iron and steel making, non-ferrous metal smelting and refining, cement and ceramics manufacturing and fertilisers. The existence of a domestic smelting and refining industry encouraged an aggressive search for minerals from overseas as domestic mine production proved incapable of meeting growing demand. Policy at that time was oriented to assist the domestic mining industry and to protect the domestic processing industry.

In non-ferrous mining, Japan was in 1988 among the world's 10 biggest producers of zinc ore and concentrates, with about 3 per cent of Western World production. In recent years, production has been declining. Japan is also the second largest producer of refined gallium, indium and refined zinc, and the third largest producer of refined copper, refined nickel and titanium sponge. Japan is also the world's leading producer of minor metals including bismuth, cadmium, germanium, electrolytic manganese dioxide, selenium and tellurium. The country remains the world's largest importer of, among others, copper concentrates, iron ore, nickel ore and phosphate rock.

Since 1982, Japan's mining industry has suffered from fast depleting ore reserves, low metal prices, high production costs and increasing substitution of industrial materials. In 1986, because of a drastic decline in domestic market prices of most non-ferrous metals and lower import prices of minerals and metals resulting mainly from the rapid appreciation of the Japanese yen, the mining industry went into a state of deep recession.

Because of world-wide excess capacity and depressed prices, the Japanese minerals and metals industry have undergone several rationalisation and restructuring programmes, the most dramatic example being the aluminium industry. In the context of the Depressed Industries Law of 1978, and the Structurally-Depressed Industries Law of 1983, the capacity of the primary aluminium was severely reduced from 1.6 million tonnes to virtually nothing. These restructuring measures were extensive in nature and implemented rapidly.

Other non-ferrous metals smelting and refining industries have been able to maintain capacity because of their role in providing raw materials to the steel industry or to the fast-growing high-technology sector. Nevertheless, the Japanese private companies in a designated declining industry bear an unusually large share of the burden of adjustment. The firms, rather than the public agencies, play a leading role in relocating redundant workers, and in phasing out excess capacity. To enable firms to administer and partially finance structural adjustment, Japanese public policy aims to increase the resources available by directly co-ordinating industry-wide capacity reduction or by permitting cartelization for this purpose.

In addition to government bodies with general responsibilities affecting the minerals and metals industries, such as the Finance Ministry (conducting tax and exchange rate policies), the Foreign Affairs Ministry (dealing with co-operation matters) or the Environment Agency, the Ministry of International Trade and Industry's (MITI) scope of actions covers a wide range of economic aspects, such as trade policy, the regulation of foreign investment and technical co-operation.

There are six major agencies with responsibilities in the non-fuel mineral policy area. Within MITI, three are concerned with financing overseas mineral resource development and resource flows. They are:

i) Overseas Mineral Resources Development Corporation (OMRD). Created under MITI's initiative, OMRD is a joint venture of the Japanese Government (50 per cent) and a consortium of most of the Japanese minerals and metals companies.

ii) MITI's Mining Division of the Agency of Natural Resources and Energy (ANRE). The ANRE is in charge of the rational development of mineral resources and the securing of stable supply of energy to Japan. Its Mining Division has jurisdiction over metallic and non-metallic minerals, as well as non-ferrous metals (excluding light metals) and over the Metal Mining Agency of Japan, but has no direct operating responsibilities.

iii) Metal Mining Agency of Japan (MMAJ). Under the Metal Mining Agency Laws of 1963, the MMAJ was created under the supervision of MITI to execute and promote, in close collaboration with ANRE, government mining policy for non-ferrous metallic minerals. MMAJ's mission is to implement exploration and exploitation programmes in Japan and overseas, to finance non-ferrous metals stockpiling, to finance and administer rare metals stockpiling, to develop new technology for mineral exploration and extraction, and to provide financial assistance to mining companies to prevent mining pollution. MMAJ is also authorised under the supervision of MITI to ensure that Japan's mineral policy is followed.

In addition, the Export-Import Bank of Japan (EXIM) extends loans and yen credits to purchase Japanese goods and services for natural resource development. The EXIM grants its loans and credits to companies, Japanese or not, that commit to sell the resulting production to Japanese markets. In addition, the Bank extends import credits to Japanese importers of "vital resources". Under its exports, imports and overseas credit programmes, the EXIM not only opens markets for Japanese-produced goods and technical services, but also secures natural resource supplies to the Japanese economy.

Two agencies involved in mining relations with developing countries are:

i) the Overseas Economic Co-operation Fund (OECF) finances mining exploration and development overseas;

ii) the Japan International Co-operation Agency (JICA) was established in 1974 to provide technical assistance and financing to developing countries. The JICA conducts exploratory surveys and feasibility studies and extends loans and grants for the development of a natural resource-related infrastructure.

A consultation procedure between the government and the industry exists in Japan. A large number of councils covering specific subjects have been set up as advisory bodies to the government, such as the Mining, the Aluminium and the Iron and Steel Industry Committees. These councils — composed of civil servants, leaders from the industries and industry associations concerned, experts, scientists, general consumers and leaders in financial circles — investigate and deliberate at the Minister's request or on their own initiative on future policies in their respective administrative areas. Their reports constitute a very important element in MITI's planning and design of actual measures and policies. In fact, they play a significant role in the consensus building.

Japan's economic security relies on the stability of its foreign resource supplies. Redefining the national goal as "comprehensive security" explicitly to include mineral security, the main objectives of the country's mineral policy, as defined in the White Book on Raw Materials (published by MITI in October 1971) and the White Book on Co-operation (published in January 1972), are the following: (i) secure stable sources of minerals; (ii) systematically develop domestic mineral resources; (iii) actively promote development of overseas mineral resources through economic co-operation with mineral-rich developing countries; and (iv) stockpile rare metals.

A new set of discussions involving the Japanese Government and the private companies is currently underway, aimed at achieving a consensus on a new policy towards the non-ferrous metals industry.

PROSPECTING AND EXPLORATION

In order to promote the systematic development of domestic resources, exploration by the so-called "three stages method" is carried out according to the long-term programme of domestic exploration. The programme was planned and brought into effect by the government in the early 1960s with the aim of increasing the domestic ore reserves of non-ferrous metals. It has been specified under two successive "Long-Term Domestic Prospecting Plans". The first started in 1966 in

28 districts while the second began in 1973 in 25 districts. In 1989, 19 domestic districts were newly added as investigation areas for domestic exploration.

Geological services

To develop domestic mineral resources, the government since 1963 has conducted regional and detailed geological surveys through the MMAJ. These surveys, which constitute the first of the three stages, are intended to elucidate the outline of geological structure in areas where the exploration of mineral resources needs to be promoted. Annual work plans are drawn up by the MMAJ, which then supervises work carried out in co-operation with university scientists and engineers, mining firms and prefectural governments. Main targets are copper, lead, zinc, manganese, gold and tungsten.

In addition, in view of the importance of securing rare metal mineral resources, a special survey on domestic rare metal mineral resources was introduced in 1985.

At the second stage, the detailed geological survey is carried out in the most likely areas selected by the preceding stage. This is intended to elucidate the ore-bearing geological structures in target areas of high potential by means of systematic core drilling and to suggest to companies the places which seem to reserve ore deposits. MMAJ carries out this survey by receiving subsidies and expenses from the national government (2/3), the prefectural governments (2/15) and vested private companies (3/15). Main targets are copper, lead, zinc, manganese, gold and tungsten.

Lands availability

The exploration and mining activities are in fact restricted in natural parks and preservation areas of natural environment. The percentage of such areas restricted to exploitation and mining activities is not available. Article 2 of the Mining Law stipulates that the state fulfils the function of providing the right of exploration and acquisition over the mines which are not mined. In other words, the state owns minerals which are not mined, as well as holding the mining rights for such minerals.

Granting exploration permits and concessions

The basic law affecting minerals in Japan is Mining Law 289 of 1950. Article 17 of the Mining Law limits exploration and mining rights to Japanese nationals and Japanese corporations. However, if a treaty decides otherwise, foreign legal persons have the possibility of obtaining such rights. There is no such case to date. Granting of mining rights and mining leases, as well as enforcing mining safety, is administered by MITI's regional bureaus.

Exploration incentives

The third-stage of the three stage process is the actual exploration by private companies. In the case of small-sized companies, 50 per cent of exploration costs of new deposits are given as subsidies by the government. Also, research and development (R&D) activities on by-product and technical training for mine workers are subsidised to support small-scale mining. This programme, in operation since before World War II, was previously conducted by the Industry Bureau of the MITI. MMAJ also encourages the large private sector companies to delineate new deposits by providing government loan subsidies for up to 80 per cent of the costs in remote virgin areas and up to 70 per cent in other areas. Repayment period is 12 years (three year grace period) for remote areas and seven years (two year grace) for other areas. At the beginning of 1989, the interest rate was 6.15 per cent per year. Main targets are copper, lead, zinc, manganese, gold, molybdenum and tungsten.

MINING AND METALS PRODUCTION

Direct and indirect support

Since 1978 the MMAJ has provided metal mining companies with low interest loans for stabilising the metal mining industry. Loans are made from a base fund borrowed from the Industrial Bank of Japan and private financial institutions under government guarantee. The system of the Emergency Loans Fund for the metal mining industry is applied when market prices for metals are below specified floor prices. On the other hand, when market prices rise above previously determined ceiling prices, private companies have to repay the loans as well as the interest, according to a specified ratio.

Thus, in late 1978, an emergency scheme was introduced for the 1978-80 period to bail out copper and zinc mines that had been weakened by the post-1975 recession in metal prices. Through the Metal Mining Industry Emergency Loan Fund, soft loans at very low interest rates became available when market prices dropped below defined floors. As with most Japanese schemes for loan support, the government's commitment is limited to guaranteeing the MMAJ's borrowings from commercial banks, and to making up the difference between the MMAJ's commercial borrowing rate and its subsidised lending rate. In 1986, gold and tungsten were included in the scheme, and Japanese non-ferrous mining companies applied again for the emergency fund to cope with declining zinc and copper prices.

In contrast, the energy tarification policy of the Japanese Government is rather strict towards the non-ferrous metals industry. In effect, energy tariffs are set taking into account total production costs plus a normal return on investment and there are no special programmes to serve big consumers under long-term purchase contracts or on a preferential basis. On the contrary, tariffs rise proportionally with energy consumption in order to force big consumers to undertake energy conservation investments.

The Japanese Government is engaged in a wide range of energy programmes which have an impact on the energy supply picture of the minerals and metals industry. Thus, for instance, in an industry-wide energy conservation programme to convert electricity-generating plants from oil to coal, low-interest loans were granted to several companies, notably in the primary aluminium sector.

Grants and low-interest loans were provided by the Japan Development Bank to energy plant owned in association by aluminium and energy producers. Temporary special tariffs were accorded to primary producers engaged in the conversion of their energy sources from oil to coal; specific grants could cover up to 15 per cent of the total investment expenses. Low-interest loans were also provided by government long-term development banks to finance the construction of new coal-fired power stations.

Structural adjustment policy

The MITI has on successive occasions "guided" the minerals and metals industry either to "orderly" expansion to avoid excessive competition or to capacity reductions and restructuring in order to increase the industry's competitiveness. According to Upham[1], this is so because group identification in Japan is so intensive that it inevitably leads to virulent excessive competition, rather than collusion, among groups. Therefore, this justifies government intervention to moderate competition in Japan, not a policy that encourages its excesses.

On the other hand, to enable firms to assume the burden of administering and partially financing structural adjustment when changes in technology or in international factor prices have rendered Japanese industries uncompetitive in world markets, Japanese public policy has on occasion co-ordinated industry-wide capacity reduction or allow a certain degree of cartelization for this purpose. Even where explicit agreement on prices or production levels is prohibited, a planning process or legal capacity-reduction cartels may be expected to restrain the tendencies to engage in drastic price-cutting.

The MITI could at times exercise an influence over the industry, through, for example, its power to authorise "depression cartels" as circumstances warrant. It has at times used informal cartel and merger policy to foster growth in emerging sectors, to counteract cyclical downturns in mature industries (recession cartels).

By the late 1970s, however, the increasing competition from the newly industrialising countries and the dramatic rise in energy costs contributed to put many of Japan's hitherto successful industries at an international competitive disadvantage. In 1978, the combined pressures of the first oil shock, the yen's appreciation and volatility of metal prices led to severe financial difficulties in the domestic minerals and metals industry.

That year, the "Depressed Industries Law" (DIL) allowed threatened firms in structurally depressed sectors to qualify for preferential loans drawn from the newly created Depressed Industries Credit Fund, as part of their restructuring programmes (cartelization of selected structurally depressed industries). This Law aimed at the elimination of obsolete and inefficient capacity of specific industries by means of the constitution of cartels which were legally voluntary. Those firms that chose not to participate (the "outsiders"), were not legally constrained by cartel agreements or MITI recommendations. Aluminium smelting, electric furnace steel and ferro-silicon were among the 14 industries considered as structurally depressed in 1978.

These temporary measures for stabilisation of specific depressed industries is only one element of Japanese public policy towards its declining industries. Some financial incentives in the form of loans and loan guarantees were also provided. They included:

- low-interest loans for equipment to facilitate the transfer of a firm to a new line of business;
- low-interest loans to finance retirement allowances to individuals made redundant by capacity reductions;
- guarantees for new loans from private banks to pay-off existing loans on plants and equipment, permitting the excess capacity to be scrapped.

These loans were provided by the Japan Development Bank, a quasi-government corporation. The amount of loans and loan-guarantees under the 1978 Law totalled US$148 million.

There were also laws protecting workers and small business enterprises. Temporary measures for the unemployed provide subsidies to help the firms to adjust their labour force to the new market conditions. The subsidy can amount to as much as full salary for six months if the Ministry of Labour determines that a worker's new employment can be regarded as training. A second category consists of special allowances related to early retirement (typically between 55 and 60 years of age in large companies), often paid in a lump-sum bonus of three or four times their annual wage rather than a continuing pension.

Employment adjustment is also facilitated by the purely private practice of placing discharged permanent workers in jobs with other firms in the same business group.

The second oil shock plunged the above-mentioned industries into a even deeper depression. In 1983, the DIL, was converted to the "Structurally-Depressed Industries Law" (SDIL), and the three industries previously designated remained under the protection of these measures. The means of MITI's intervention were correspondingly expanded from the orderly disposal of capacity to the provision of appropriate means of production. In both the production and price cartels, MITI has largely delegated control to the trade associations concerned and the companies themselves.

The plans jointly prepared by the MITI and the industry, known as the Basic Structural Reform Plans, provided for:

i) joint production, marketing, purchasing, and storage activities;

ii) shifts in production among firms, including agreements to specialise in certain products or to merge management and production in whole or part;

iii) investment in energy conservation or fuel conversion and in increased production efficiency and product quality; and

iv) the development of new products and technologies.

To facilitate these goals, the SDIL provides a broad range of new tax incentives to complement the loan guarantees already available, including grants for research and development, special depreciation for modernisation and capacity-reduction investment and the extension of the period for carry-forward deductions from five to ten years.

The case of the primary aluminium industry

The aluminium industry suffered severely from the 1973 and 1979 energy shocks as most of the electricity consumed was generated in oil-fired plants. Moreover, Japanese primary producers had no long-term contracts for electric power to moderate the impact of rising electric prices.

In response to the Japanese companies' difficulties, MITI has played a key role in restructuring the industry through the implementation of several measures:

i) the promotion and co-ordination of the progressive reduction of excess capacity;

ii) the promotion of the development-for-import of overseas primary aluminium projects;

iii) the promotion of technological research and development;

iv) the stabilisation of employment.

In 1978, through a consensus between MITI and the industry, the excess primary aluminium capacity to be scrapped was identified at 500 000 tonnes, to result in a reduction of capacity from 1.6 million tonnes per year (tpy) to 1.1 million tpy. To reach this ambitious goal, a tariff-quota system was applied in 1978 and 1979. The Japanese primary producers and fabricators were allowed to import aluminium ingot at reduced tariffs on a first-come, first-served basis until the total amount of these imports equaled the amount of the demand in excess of the capacity that was to be retained. Once this limit or quota was reached, importers were again subject to normal tariffs.

In turn, their tariff savings (less a handling fee) were turned over to the Association for the Promotion of the Structural Improvement of Aluminium Industry. This association then made payments to primary producers of 6.6 per cent of the book value of the capacity that the firm had scrapped or mothballed. The subsidy served to cover the interest cost associated with capacity that was no longer in operation.

After the second oil-shock, it became clear that the goal to retain a 1.1 million tpy capacity was unrealistic. In 1982, this goal was reduced to 700 000 tpy and the tariff-quota system was replaced by a tariff-exemption system that allowed primary producers to import aluminium up to the amount of the capacity they had closed duty free. Producers could benefit from the tariff which was about 9.0 per cent of the import price. The amount of the subsidy under both systems is estimated to have totalled 28 billion yen (US$112 million) through 1984. In addition the government financed the Light Metals Stockpiling Association which bought 3.7 billion yen (US$14.8 million) of ingot from 1976 to 1983.

The aluminium crisis made the government change its mind again in 1984, when it set a new target of 350 000 tonnes of excess capacity to be scrapped. In this context, the Ministry of Finance allowed the companies to import 350 000 tpy of aluminium ingot in the 1985-1987 period with a tariff of only 1 per cent. In the course of 1987, the normal 9 per cent rate for the remaining imports was scaled-down to the same level.

The government assisted the industry in carrying out basic research on new smelting processes through the Research Association on the New Smelting Process for Aluminium. Public assistance was less than the losses borne by the aluminium industries which amounted to 1 105 million yen (US$4.4 million) from 1976 to 1983.

With regard to employment adjustment, the number of workers in aluminium smelting has been reduced from 15 000 in 1975 to about 4 000 in 1983 and virtually none today. About half of the workers that left the industry were reemployed by affiliated companies of the same business group.

As the reduction of most of Japan's import tariffs on metals is being discussed today, the framework of these laws could be applied to the restructuring of other non-ferrous metals industries in Japan, such as nickel, lead, zinc and copper.

Taxation

Japan's aggregate effective income tax rate for minerals and metals firms is estimated at about 51.0 per cent. This includes corporate tax (33.48 per cent), prefectural tax (1.67 per cent), municipal tax (4.12 per cent) and enterprise tax (12.0 per cent). Mining companies operating on Japanese soil are however exempted from the enterprise tax. Concessions available include accelerated depletion in the first year of expenditure for pollution control and energy saving expenditures, and an additional depreciation deduction of 20 per cent of the cost of specified machinery and equipment in the first year. Mining companies benefit also from accelerated depreciation under digressive conditions.

Other significant tax measures related to the minerals and metals industry were described above (the domestic prospecting reserve fund and the price fluctuation reserve). There are also specific tax breaks for restructuring programmes. Thus, industries ascribed permanently depressed status receive investment tax credits of 7 per cent of the cost of new equipment subject to a credit ceiling of 20 per cent of tax due. Unused credits can be carried over for 1 year.

Since 1965, an allowance for natural resources exhaustion, similar to that of the US "depletion allowance" or "provision pour la reconstitution du gisement" in France, has been in operation and will be applicable until 1995. As a result, Japanese mining corporations can deduct, according to each exploration and mining investment, 13 per cent of the total selling price of the mines or 50 per cent of the total income from the taxable revenues. The benefit gained from this allowance must be invested in exploration and mining activities within a three year period.

HORIZONTAL POLICIES

Environment

One of the crucial aspects of Japan's mining policy is the prevention of heavy metal dispersion around mining and smelting facilities and the restoration of contaminated farm lands. The national pollution control effort began in 1967 with the Basic Law for Environment Pollution Control, and was strengthened by subsequent legislation, including a basic law on air pollution control in 1968.

Controls of emissions in the atmosphere are set both by the Environment Agency, in the context of the Air Pollution Control Law, and Prefectural Governments. Maximum levels of lead in emissions are: (i) 10 mg/m^3 in the case of roaster, converter, smelting furnace and dryer for copper, lead or zinc smelting; (ii) 30 mg/m^3 for sintering machine and blast furnace for copper, lead or zinc smelting. Stricter standards of 1.5-7.0 mg/m^3 may be settled by Prefectural Ordinance. There are no specific regulations for maximum levels of non-ferrous metals in the general atmosphere.

The Water Pollution Control Law issued by the Environment Agency fixed in 1971 maximum lead levels in industrial effluents at 1.0 mg per litre. Stricter standards of 0.1-0.8 mg/l may be set by Prefectural Ordinances. On the other hand, the national environment standard for health is set at 0.1 mg/l of lead in water.

The regulation of lead in air in the atmosphere of working area of industrial works is made by the Ministry of Labour in the context of the ordinance on the prevention of lead poisoning of the Industrial safety and health Law. Settled in 1972, maximum levels for lead in atmosphere are 0.1 mg/m^3.

In order to meet stronger regulations and to ensure their effectiveness, the minerals and metals industry has had to make positive efforts to develop anti-pollution techniques and to install anti-pollution equipment satisfying these regulations and standards. In order to advance private industry's anti-pollution technological development and equipment installation, the government provided information, offered guidance, has done some technical development work on its own and implemented fiscal and financial support measures.

Most of the regulations affecting minerals, such as the Mining Safety Law, the Law on Extraordinary Measures to Deal with Mine Pollution and the Basic Law for Environmental Pollution Control, are administered and enforced by the MITI, through various departments and divisions under the Industrial Location and Environmental Protection Bureau as well as its regional bureaus.

MMAJ is responsible for the administration of several programmes to deal with the pollution problems associated with mining and to ensure that mining operations respond to national pollution control. These programmes consist of financial assistance to firms for control of pollution from mining; administration of an environmental protection fund for pollution control work as mines close; pollution control engineering for abandoned mines; research and development for mine pollution control; and technical surveys and guidance for prefectural governments for mine pollution control.

MMAJ supplies loans to mining companies who are obliged to carry out anti-pollution works (since 1973) and to bear the cost of restoring farm lands (since 1975), as required by the anti-pollution laws and to prevent pollution arising from mining and smelting facilities. The Agency also guarantees the liabilities of mining companies on their loans from city banks for their anti-pollution works.

The maximum amount for a loan for anti-pollution facilities is 70 per cent of the total expenditure for large companies and 80 per cent for small and medium scale ones. Annual interest rates are, in the case of large companies, 5.3 per cent for anti-pollution facilities and 7 per cent for farm land restoration. In the case of small companies these rates are respectively 3.5 per cent and 7 per cent. The repayment term is less than 15 years including two years grace in both cases.

MMAJ also supplies loans to mining companies to pay for preventive constructions of mining related pollution, which have been undertaken by prefectural governments on farm land that was affected by specific noxious materials (cadmium, copper, arsenic), pursuant to the anti-pollution laws.

In addition, MMAJ administers the Environment Protection Fund. The Agency receives and manages the obligatory deposits moneys which are contributed, pursuant to the Law on Extraordinary Measures to deal with Mine Pollution, by the mining companies of active mines, and which are to be withdrawn for exclusive use in the construction of anti-pollution facilities at the time of the closing down of the mines. The amount of money of deposit money of each mine is determined annually by the nine Regional Mine Safety and Inspection Bureaus of the Government.

Prefectural governments are charged by anti-pollution laws with responsibility for abatement of pollution from abandoned mines. Local offices of the MMAJ provide engineering assistance to Prefectural governments for these activities. Technical fees are borne by the Prefectural governments, while operational expenses are borne by the MMAJ.

Because the overseas location of resource processing industries is thought desirable from the long-term perspective, environmental considerations have also led to the promotion of industrial relocation plans in mineral rich developing countries.

Research and development (R&D)

Since 1975, the MMAJ has been actively involved in the development of new technologies in the fields of mineral exploration and extraction. Projects include:

- The Mineral Resources Centre Information gathering, translating and analysing information from around the world on mining, exploration and geology, for use by both the public and private sectors.
- Development of new exploration methods for discoveries in more complicated structures, and in less accessible areas, such as electronic prospecting systems, remote sensing analysis and prospecting systems for rare metals.
- Technology for the recovery of rare earths, in order to secure the stable supply of inputs to super-conductive materials.
- Deep-sea mineral resources survey: in the early 1970s, as the government adopted ocean development as one of its priorities, marine minerals' exploration has been included in MMAJ's activities. Projects include governmental exploration for manganese nodules, carried since 1975 in research vessels, and for polymetallic sulfide deposits of the East Pacific Rise, as well as the technical development of bottom siting marine drills.
- Development of a Deep Seabed Mining System: MMAJ has been involved in the National Research and Development programme (large-scale project) on manganese nodules directed by the MITI. The Technology Research Association was established for this project by 20 companies in 1981.

In order to solve mining-related pollution problems various technological investigation programmes have also been undertaken since 1974. Efforts focus on mine drainage and tailings and water.

As far as metallurgical R&D is concerned, it is important to note that most of the Japanese metals companies belong to larger conglomerates (the "keiretsu") and thus have access to their technical competence and the result of their investments on technological research. On the other hand, as a great part of their customers lie within the same group, technical co-operation for the development of new products in accordance to the consumer's needs also becomes easier.

Industrial R&D financing by the government is virtually confined to special programmes addressed to strategic sectors. That is the case, for instance, of the research programme on the development of new materials, that in its high purity metals sub-programme corresponds to the new strategic orientation of the Japanese minerals and metals companies to move downstream to more value-added products.

Even in the context of these special programmes, funds available are relatively limited. In contrast, private companies play a much more active role. The results of government sponsored R&D in Japan are not proprietary to the firms taking part in the research programmes but belong to the government. Licenses may be sold to all interested firms and receipts from such sale revert to the government.

Foreign investment controls

Foreign investment in mining industry in Japan is subject to the Foreign Exchange and Foreign Trade Control Law. This law is applied to cases where foreign investment covers more than 50 per cent of the capital of the newly created mining corporation. On the other hand, there is no control over foreign investment in metals industries in Japan.

Trade policy and marketing

It is interesting to note that the steel industry's purchases of iron ore and coal are grouped in centralised procurement agencies that buy the quantities needed for all the members at the same conditions. This procedure reinforces the bargaining power of the industry vis-a-vis its suppliers and also contributes to attenuate competition within the industry. This sort of institution does not exist in the case of the non-ferrous metals sector. However, most of the ore purchases of the Japanese industry are made in the context of long-term contracts entered into by special consortia, where several metals companies are regrouped. The purchasing agreements contracted under these consortia (and the financing contracts permitted by them) have often been key elements in the development of mining projects in developing countries.

Import tariffs on wrought products remain highest on aluminium semi-manufactures (9.2-10.2 per cent) and are lowest on tin semis (3-3.9 per cent). In the case of copper, lead and zinc, Japan applies a sliding tariff scale, whereby the applied rate varies with the price of metal; when the metal price is above a reference price, the applied tariff per kilogram diminishes. The applied tariffs do not exceed the bound tariff levels[2]. GSP rates are all zero, but ceiling quotas apply on all products under study except for zinc, other than unwrought.

Commercial policy considerations deserve a special mention in the case of copper. Japan has extensive copper smelting and refining capacity: production topped the 1 million tonne mark in 1989 for smelter output and was just below for refined copper (990 Kt). Japanese smelters produce at a comparable level to that of the two largest copper mining countries, Chile and the United States, although domestic mines now output less than 1 per cent of these latter two (15 Kt compared to 1 500 Kt). This makes Japan far and away the largest importer of copper concentrates — taking some 60 per cent of world exports of copper concentrates. Japan maintains free entry for concentrates and matte; on blister and refined copper, bound tariff rates are 21 yen/kg. However, under the Japanese sliding tariff system, the specific rate decreases with the increase in the price of refined copper and above a certain price, duties become free. Conversely, when copper prices fall, the tariff is reinstated and can increase back to the mentioned bound level. This has at times led to a situation of keen competition for concentrates with other custom smelters, mainly in Europe. Following a joint request by the European Communities and Japan for conciliation through the Director-General of GATT, a good offices report found that the Japanese tariff on refined copper did not violate the terms of Japan's obligations under the GATT, but the advisory opinion recommended that the parties negotiate on a reciprocal and mutually advantageous basis to resolve the long-standing dispute.

The Export Control Act requires export approval for copper and the current practice is to grant automatic approval for export requests. Before 1979, such procedure had been enforced in the case of copper, copper alloy, semi-manufactured metal and scraps.

CONSUMPTION/SUPPLY

Securing overseas supply

The current policy concerning foreign resource supplies contrasts sharply with Japan's pre-war policy. During the 1930s and early 1940s, Japan strived for direct control over foreign resources. Following the country's economic boom in the 1960s and the 1970s, Japan developed a policy to secure mineral supplies based on develop-for-import schemes.

The flexibility of the arrangements enables the involved parties to fit their structure to suit their specific needs within the nature of each particular venture. The three basic types of arrangements consist of the following structures:

i) Production sharing joint-ventures are financed by a consortium, or joint-venture, of foreign and host-nation investors. In return for supplied equity, each partner receives a specified share of the project's output. Some recent examples include the Albras Aluminium project in Brazil and the Rio Tuba Nickel project in the Philippines.

ii) Buy-back arrangements are similar to production sharing, except that the foreign investor provides loans instead of investment capital. Export credits and subsidised loans are extended to the producer nation to facilitate purchases of machinery and other equipment needed for the resource development project. The resulting output is committed to the creditor as repayment in kind for the credits or loans. Host nations can thus develop their resources while maintaining local ownership and Japan secures its resource supply without risking nationalisation of its overseas investment. An example of a buy-back agreement is the development of a copper mine in Peru by Mineroperu, where Marubeni secured 60 million tpy of concentrates (about 70 per cent of the output) through a loan of US$25 million.

iii) Technical assistance arrangements, like buy-backs, are made on the basis of extending credits and loans to host nations. Technical services, exploration techniques, mining methods, processing technologies and management skills are purchased. An example of a technical assistance agreement is the development of the Mamut copper mine in Malaysia, where OMRD took a 51 per cent share of the capital and secured 100 per cent of output.

A common feature of all these operations is that Japanese firms often organise syndicated loans or investment consortia. However, the group approach is more extensively adopted in buy-back and technical assistance arrangements than in production sharing joint ventures. In effect, the quasi-market nature of the first two types of interventions intrinsically leave the Japanese buyers with less bargaining power and less control over flows of output than the equity investment strategy.

Japanese authorities administer their overseas exploration and development assistance programmes so as to diversify sources of supply of minerals and metals. The aim is to increase the number of countries supplying a particular mineral to Japan and to diversify sources among the greatest number of countries.

Support of exploration

There are three regular programmes for overseas mineral exploration run by the MMAJ, and a further special programme for mineral exploration in China. The three regular programmes are: subsidies to Japanese mining companies for geological surveys; subsidies to Japanese companies for joint exploration projects with foreign partners; and financing for overseas exploration. In addition, Japanese businesses typically receive tax breaks for the costs of establishing an overseas subsidiary or developing an overseas market.

Since 1968, MMAJ has provided subsidies for expenditures on geological exploration and exploratory drilling undertaken by Japanese companies in foreign countries. MMAJ also provides equity participation to Japanese mining companies for overseas exploration or loans in exploration undertaken by Japanese private companies in foreign countries. Equity participation is made through the OMRD. Its investment is limited to 50 per cent of total expenditure (70 per cent in special cases) and covers all countries. If the venture proves successful, MMAJ policy is to sell stock to private Japanese minerals and metals firms "at reasonable prices".

Provision of loan finance is also limited to 50 per cent (70 per cent in special cases) but covers all countries only in the case of rare metals. For all other non-fuel minerals such OECF loans of up to 70 per cent of the exploration costs apply only for overseas exploration in developing countries. The repayments are up to 15 years including a five year grace period. The repayment-after-success system was introduced in 1985 for rare metals deposits.

MMAJ also owns a research vessel that participates in the investigation of deep ocean floor manganese nodules, seabed hydrothermal deposits and cobalt-rich crusts. Historically, the equity participation has been directed towards copper resources and manganese nodules.

Foreign assistance to producers

MMAJ engages in two types of technical assistance related to mineral resource development in less developed countries: geological surveys and regional development planning.

In response to requests from the governments of developing countries, MMAJ participates, jointly with the Japan International Co-operation Agency (JICA), in basic surveys of mineral resources, such as topographical and geological mapping, within the context of specific projects discussed and agreed between the governments concerned. Thus, for example, MMAJ has a bilateral agreement for prospecting in China. In 1987, MMAJ reached an agreement with the government of China for joint exploration and feasibility studies for 15 metals over a period of five years. Minerals covered are chromium, cobalt, columbium, copper, hafnium, lead, molybdenum, nickel, rare earths, silver, tantalum, titanium, vanadium and zirconium.

Overall, from the beginning of this programme in 1970 to 1988, 83 projects were carried out in 31 countries in Africa, the Middle East, Asia and on the American continent.

Co-ordination of bilateral aid with mineral-rich countries has been reinforced especially where the domestic minerals and metals industry was losing its competitiveness. These undertakings were facilitated by the fact that Japanese business is accustomed to taking an economy-wide view and by government-to-government negotiations.

For example, as part of the "Fukuda Doctrine" announced in 1978, Japan increased its aid to ASEAN countries to include financial assistance to some major industrial development projects. Examples include the Asahan aluminium smelter in Indonesia and the Albras aluminium project located in the Amazon region of Brazil. Both these projects received the designation of "national project", meaning that they are of special importance for the countries concerned and were negotiated at government level. They were part of a scheme of "natural resource alliances" negotiated with these countries. Because of the appreciation of the yen and low US dollar-based prices for primary aluminium in the world market, in 1986 Albras and Asahan Aluminium received financial assistance from the Japanese Government through the OECF. Both projects had suffered heavy losses and experienced difficulties in repaying interest and principal on their yen-denominated loans.

On the other hand, since 1984, MMAJ has co-operated with developing countries in providing mining information under the Official Development Assistance scheme. Mining information on markets and technical assistance are thus offered. A research and development co-operation project in the field of separation technology of complex ore minerals in developing countries has also started recently.

Export cover and guarantees

Mining projects overseas also qualify for International Trade and Insurance Account funding and other facilities, under MITI.

Insurance to domestic companies investing overseas

The Japanese Government promotes private investment in overseas mineral resources by guaranteeing loans made for such purpose, providing political risk insurance and supporting mineral projects.

MMAJ guarantees the liabilities of Japanese companies receiving loans from Japanese banks (including the EXIM) for development projects in all countries up to 80 per cent. Nine projects have been guaranteed from 1968 until 1986, for a total of 34 900 million yen, including the PT Inco and Rio Tuba nickel projects in Indonesia and the Mamut copper mine in Malaysia.

Japanese minerals and metals companies investing overseas also qualify for Japan's general limited political risk insurance programme, which is administered by the EXIM. Guarantees cover risks of war, nationalisation and exchange rate and are extended for investments in designated "safe" countries. They include not only the initial investment, but also reinvested profits and repatriated dividends, up to 10 per cent per year. Applicable compensation is up to 90 per cent of the net residual value of the assets; coverage is for 15 years.

Direct intervention in foreign projects

OMRD and OECF have facilitated consortium formations by providing investment capital and by becoming equity shareholders in some ventures, like the majority share held by OMRD in the Mamut copper mine in Malaysia. On the other hand, OECF holds 40 per cent of the 49 per cent share held by the Japanese consortium that invested in the Albras aluminium smelter in Brazil.

The OECF also invested in the Asahan aluminium smelter in Indonesia where it holds 50 per cent of the 75 per cent share controlled by a Japanese consortium composed of a group of private firms, including five major aluminium companies and seven sogo shosha. The JICA financed development of the port facilities, highways and other components of the infrastructure and the EXIM extended long-term loans for equipment purchases.

Stockpiles

Rare metals stockpile

MMAJ was authorised in 1983 to finance and administer stockpiling of rare metals (cobalt, chromium, manganese, molybdenum, nickel, tungsten, and vanadium) vital to the Japanese iron and steel, machinery, electronic, and high-technology sectors, for the equivalent of 60 days consumption. Until 1987, three different programmes were used: National Stockpile, Private Stockpile and Joint Stockpile, with government and private sector participation. In 1987, in order to reduce the government expenditure and make it easier for private companies to participate, the scheme was changed resulting in abolishing the Joint Stockpile.

a) National Stockpile: this scheme is managed by MMAJ and it is intended to maintain national economic stability in the event of medium and long term supply disruptions. Its allocation is 42 days or 70 per cent of the overall target.

b) Private Stockpile: this scheme is on a voluntary basis by the private sector and managed by the Japan Rare Metals Association. It is aimed at various possible short term supply disruptions. Its allocation is 18 days or 30 per cent of the overall target for the stockpile.

At the end of fiscal year 1990 the stockpile had achieved stocks equivalent to 35.5 days on the basis of 1986 consumption. Government guarantees for the loan used for the purchase of materials were worth 34 billion yen, subsidy for interest and administration costs some 10.5 billion yen and warehouse costs for the National Stockpile, 1.6 billion yen.

Base metals stockpile

In order to stabilise the metals markets and secure long-term sales contracts the government started to stockpile non-ferrous metals in 1976, following a downturn in markets for the various metals which left inventories at uncomfortably high levels. The stockpile's main objective has been to ensure that Japanese commitments for long-term purchase contracts were met and thus contribute to the stabilisation of export earnings by Japan's major suppliers, which was seen as a major factor for their economic development.

The stockpile loan fund is managed by the MMAJ. The fund, financed from Japanese city banks at the long-term prime rate, is guaranteed by the government.

Practical stockpiling operations have been handled by two stockpiling associations: for copper, lead and zinc, the Japan Metal Stockpile Foundation and for aluminium, the Light Metals Stockpiling Association. The latter was dissolved in July 1987. Both have used loans from by the MMAJ. The Associations purchased (released) a specific amount of metal ingots from (to) the smelters at given levels of stocks in the market, as decided by the government. The ingots thus purchased are put in pledge to the MMAJ. When disposed of, the pledge is released and the loan equivalent to the amount of metal disposed of is reimbursed to the Agency. No base metals have been stockpiled since 1986.

Recycling

Recycling is rather developed for copper, lead and nickel. Since October 1991, the Law on the utilisation of renewable resources came in force, in order to promote the recycling of non-ferrous metals and others. As a result, the recycling of aluminium has been promoted through separate collection of used aluminium cans.

NOTES AND REFERENCES

1. An example of this school of thought is, for instance, Upham (1986).
2. Japan applies a sliding tariff scale in which: the rate of duty is 8 yen/kg for refined zinc valued at not more than 242 yen/kg (customs value); duty free for refined zinc valued at more than 250 yen/kg; and a rate of duty equal to (250 minus the customs value) when the customs value in more than 242 but less than 250 yen/kg. The bound rate of duty is 8 yen/kg for refined zinc. A similar sliding scale applies for refined copper in which: the rate of duty is 15 yen/kg for refined copper valued at no more than 485 yen/kg; duty free for refined copper valued at more than 500 yen/kg; and a rate of duty equal to (500 minus the customs value) when the customs value is more than 485 but less than 500 yen/kg. The bound rate of duty is 21 yen/kg for refined copper.

BIBLIOGRAPHY

CROWSON, Ph. (1983), "Non-fuel mineral procurement policies", *Japan's Economic Security*, AKAO, N. (ed.), Gower, Hampshire.

GOTO, A. (1988), "The Aluminium Industry in Japan", *The World Aluminium Industry in an Era of Changing Energy Prices.* PECK, M.J. (ed.), Resources for the Future, Washington, D.C.

KOLENDA, T. E. (1985), "Japan's Develop-for-Import Policy", *Resources Policy*, December.

MITI (1988), *MITI Handbook*, Tokyo.

MMAJ, *Metal Activities of Japan*, Tokyo, Japan, several issues.

PECK, M. J., LEVIN, R. C., and GOTO, A. (1988), "Picking Losers: Public Policy Toward Declining Industries in Japan", *Government Policy Towards Industry in the United States and Japan*, SHOVEN, J. (ed.), Cambridge University Press, Cambridge, Mass.

PEMPEL, T. J. (1982), *Policy and Politics in Japan, Creative Conservatism*, Temple University Press, Philadelphia.

UPHAM, F. K. (1986), "Legal and Institutional Dynamics in Japan's Cartel Policy", *Japan's Response to Crisis and Change in World Economy*, SCHNIEGELAW, M. (ed.), E. Sharpe Inc., New York.

NORWAY

PROSPECTING AND EXPLORATION

Mining on the Norwegian mainland is covered by the Mining Act No. 70 of June 1972, Act relating to the Acquisition of Waterfalls, Mines and other Real Property etc. of 14 December 1917. The Spitsbergen Islands have separate mining legislation, entitled The Mining Ordinance for Spitsbergen.

Geological services

The Geological Survey of Norway (NGU) placed under the Ministry of Industry is responsible for the geoscience knowledge base for the Norwegian mainland. The NGU provides information and advice relating to: mineral deposits, mineral exploration, discovery and evaluation of mineral resources, engineering geology, and services for multidisciplinary scientific investigations. The NGU also co-operates with industry in carrying out prospecting for mineral deposits.

The Commissioner of Mines holds administrative responsibility for the reports filed on prospecting, exploration and exploitation of claimable minerals pursuant to the Mining Act. These reports are made public when the pre-claim and mining rights have expired.

Topographic maps and geodesic surveys are the responsibility of Statens Kartverk, Honefoss.

Lands availability

The Mining Act sets out the areas where prospecting is not allowed. In addition to these areas, a number of nature parks and other protected areas have been established during the last two decades and prospecting is prohibited in these areas.

The mining legislation distinguishes between claimable minerals and non-claimable minerals. As non-claimable minerals belong to the landowner, all prospecting on non-claimable minerals has to be arranged beforehand with him.

Exploration permits and concessions for claimable minerals

Pre-claim licences

In order to acquire rights to explore pre-claim areas for deposits of claimable minerals, as well as the priority right to obtain a claim patent in the area and to exploit the deposits, applications must be made to the Commissioner of Mines for a pre-claim licence.

Once all required information has been submitted and no provisions in the Mining Act are invoked in opposition, the Commissioner of Mines issues a pre-claim licence. Once issued, the holder may undertake whatever survey work is necessary in the pre-claim area to enable him to judge whether there is a deposit of claimable minerals rich enough to be considered commercially viable. A pre-claim area can be up to 300 000 m^2. Foreign citizens have to establish a company with headquarters in Norway in order to be awarded a pre-claim licence.

Anyone is entitled, subject to a few conditions and limitations, to a pre-claim licence to undertake exploration activities and a claim patent in respect of deposits of claimable metals on land owned by himself or by others (even without their consent). In areas which are considered "built up", on crop land, or used for public service, the prospector is to obtain special permission from the owner of the land and user. The King has the authority to decide whether areas should be withdrawn from prospecting.

The pre-claim licence permits its holder to explore for deposits of claimable minerals and gives priority for a claim patent to the area and permission to exploit the deposits. The pre-claim licensee may request a claim patent if he can argue that the richness of the deposit is of such a scale and nature that it may became commercially viable.

Claim patent certificate

Within seven years, the pre-claim holder must apply for a claim patent. To obtain the claim patent the holder has as above to make a plausible case that the deposit of claimable metal is of such a grade that it can be assumed to be, or will become within a reasonable period, commercially viable.

The exploration and extraction of alluvial gold on state-owned land may be permitted by the Ministry of Industry for up to five years at a time and for a specific area. If the deposit is commercially viable, the applicant may obtain a permit to extract gold for a 20-year period. A fee of 2 per cent of the value of all extracted gold must be paid to the Government. The state may assess the gold on the basis of international market prices.

Foreign nationals, joint-stock companies and other companies with limited liability are only entitled to explore after first obtaining the consent of the Minister of Industry. During the 1980s, all applications by foreign-controlled companies were approved by the Ministry.

Government exploration

The NGU undertakes mineral resource mapping and tries to stimulate mineral prospecting. The national mineral data base is being built up gradually, and access to it is constantly improving. The NGU runs a programme for investigation of state-owned claims, and two parallel county programmes aimed at finding mineral resources which may be exploited. The basic data on the countries are, as they are built up, made available to industry for prospecting.

The government may support exploration for amounts up to 50 per cent of expenses. This support is distributed by the Commissioner of Mines after thorough examination of applications. The total amount of this support has in recent years been NKr 4 million per year.

Taxation

In order to retain his rights under a pre-claim licence or a claim patent, the holder of the right must pay an annual fee to the Treasury (NKr 10 000 for a pre-claim and NKr 15 000 for a claim). For rights held by the state, no fee is payable.

Exploration expenditures can be deducted from taxable income as costs. The expenditures are normally fully deductible in the year of accrual, while investments must be capitalised and depreciated. There are no special depreciation rates concerning investments connected to exploration and mining. However, the cost of purchasing the mining field as such may be deducted through a reasonable depreciation rate each year, normally based on the assumed economic life of the field.

MINING AND METALS PRODUCTION

Mining

Company structure

The Norwegian (non-hydrocarbon) mining industry represents some 3-4 per cent of Norway's gross domestic product. Norway has traditionally had vast metals and mineral resources, and the mining industry was for several hundred years a very important industry. Its importance has, however, diminished over the years due to low world market prices and exhausted resources.

Table III. **Norway: extraction of non-ferrous metal concentrates (preliminary figures 1990)**

Pyrites	303 364
Copper concentrates	97 614
Zinc concentrates	34 124
Lead concentrates	5 699

Source: Central Bureau of Statistics of Norway.

Folldal Verk A/S and Grong Gruber A/S, both owned by Orkla-Borregaard and Outokumpu Oy, are two of the largest remaining mines yielding copper with zinc as a by-product. Folldal Verk is today the only producer of pyrites, which is an input in the production of sulphuric acid at Borregaard. A/S Bleikvassli Gruber, a wholly-owned subsidiary of A/S Sydvaranger, producers zinc and lead concentrates in Nordland in northern Norway. Reserves should allow production to continue until approximately 1998.

Bidjovagge Gruber A/S, a wholly owned subsidiary of Outokumpu, produced copper and gold until October 1991, when the mine was temporarily closed due to exhausted resources. A three-year exploration programme has been started in the area, but it is uncertain today whether new resources will be found which would allow the mine to be re-opened.

In 1989, the production of nickel concentrate started in Nordland in northern Norway by the company Nikkel og Olivin A/S. The company has however had several problems, leading at times to halts in production. Nickel and copper are by-products in the production of ilmenite ore at the Titania mine.

Concessions

In order to exploit minerals on a commercial basis, a concession must be obtained from the Royal Ministry of Industry. The concession is issued to a specified individual, company, corporation or foundation that must have their headquarters in Norway and an executive board whose majority, including the Chairman, are Norwegian citizens.

When a claim patent has been granted, its holder has to apply for a concession according to the provisions in Chapter II and Chapter V in Act No. 16 of 14 December 1917, before he can start mining operations on a commercial basis. The Royal Ministry of Industry issues the concession.

The Act on Mining from June 1972 distinguishes between claimable minerals, which belong to the State, and non-claimable minerals, which belong to the landowner. Claimable minerals are metals with a specific gravity of 5 or more and ores of such metals, as well as titanium and arsenic and ores of these metals and pyrrhotite and pyrites. Both nickel and bauxite are considered to be claimable minerals. Examples of non-claimable minerals are bog iron and alluvial gold. As well as granting permission for the exploration and extraction of alluvial gold on state-owned land, the Ministry of Industry prescribes the detailed conditions applicable to such permission.

Minerals which are non-claimable may be quarried by the claim patent holder insofar as this is necessary to work the claimable minerals. This is the case for example of by-products. This does not apply however if, prior to the commencement of mining operations, including trial operations, the Mining Superintendent considers that the minerals which are non-claimable occur in such quantity and form that the deposits can be worked to advantage independently.

In practice the Ministry does not grant concessions to foreigners. Foreign citizens and foreign companies have first to establish a joint stock company with headquarters in Norway and with a majority of the Board and its chairman Norwegian citizens (Act No. 16 of 14 December 1917). The concession is then granted to the company. A foreign corporation may conduct business in Norway through a subsidiary or branch. Most of the directors of the subsidiary must be Norwegian. Furthermore, authorities may require that a stipulated proportion of the capital stock be made available to Norwegians. There are also regulations concerning preference to Norwegian labour and materials, although these regulations are the same for companies owned by foreigners and those with Norwegian ownership. Mining concessions may therefore be granted to aliens and corporations on generally the same terms as to Norwegian citizens and firms.

Minerals taxation

The Norwegian tax system has been characterised by high tax rates and a narrow tax base. However, since 1 January 1992 a new tax system applies in Norway. Corporate profits are taxed at a rate of 28 per cent, instead of 50.8 per cent. Along with the reduction in tax rates, there is a considerable broadening of the tax base through a general reduction in permissible tax deductions. In general, untaxed reserves are no longer allowed. For instance, the earlier possibilities of making tax-exempt allocations to consolidation fund, classification fund etc., were removed.

In connection with the tax reform, depreciation rates are reduced. The assets are mainly depreciated according to the declining balance method. The deprecation rates are the same in the extraction industries as in other industries. The rates of depreciation (capital cost allowance) for the extractive industries are as follows:

— Office machines, etc. and acquired goodwill	30 per cent
— Articulated trucks, lorries, buses, etc.	25 per cent
— Cars, tractors and other vehicular machinery and equipment, other machines, tools, etc.	20 per cent
— Buildings and construction	5 per cent

Time-limited intellectual property that is not depreciated using the declining balance method is, as a rule, to be depreciated using the straight line method for the duration of such property rights. Non time-limited intellectual property will not normally be eligible for tax depreciation.

Research costs are normally fully deductible in the year of accrual. However, research investments can be activated and deducted through a yearly "reasonable" amount, but mainly with at least 20 per cent per year.

There are no special tax rules applying to the extractive industries, except that the extractive industries in general do not pay the investment tax. However, many other industries are also partly or wholly exempt from the investment tax.

Other fees

The mine owner is required to pay to the owner of the land an annual fee equal to a thousandth part of the value of all that is quarried from the claim patent and sold in the market. The basis of calculation for the fee is the price obtained for the products direct from the mine or any stage of metallurgical treatment. The fee has, however, an annual limit and is adjusted every 10 years on the basis of a scale laid down by the King, with special attention being paid to changes in the purchasing power of the krona. The fee is not due during the mine's first two years of operation.

Metals production

Company structure

With its extensive hydro-power resources, Norway has a large electricity-intensive production of non-ferrous metals and ferro-alloys. In 1989, value added in the Norwegian non-ferrous metals industry was NKr 8.2 billion. This accounts for 1.3 per cent of the Norwegian gross domestic product.

Hydro Aluminium A/S (HA), a wholly-owned subsidiary of Norsk Hydro A/S, is the leading aluminium producer in Norway with total capacity of 630 000 tonnes per year (tpy) in four plants plus a 50 per cent share in Sør-Norge Aluminium (50 per cent Alusuisse). HA is the largest manufacturer of extruded aluminium products in Europe (250 000 tpy), and is engaged in alumina production in Jamaica and Guyana. Elkem Aluminium A/S [50/50 Elkem A/S and Alcoa (US)] has two plants with a combined capacity of around 200 000 tpy with no company-owned down stream processing. Elkem developed the Søderberg technology in the 1920s.

Vigeland Metal Refinery A/S produces 5-6 000 tpy of super purity (6N) aluminium mostly for aerospace and electronic applications. Its share of western world production is 25-30 per cent.

Falconbridge Nikkelverk A/S in Kristiansand refines some 9 per cent of western world primary nickel (55 000 tpy). Important by-products are copper (35 000 tpy) and cobalt (2 000 tpy). It is owned by Falconbridge Inc., which in turn is owned 50/50 by Noranda (Canada) and Trelleborg (Sweden).

Norzink A/S, Odda, is owned 50/50 by RTZ (UK) and Trelleborg. Capacity is nearly 125 000 tpy of zinc.

Norsk Hydro is, together with Dow Chemicals, the world's leading producer of magnesium. Production capacity in Norway is 60 000 tpy. Hydro's new plant in Canada began production in 1990.

Table IV. **Norway: production of non-ferrous metals (preliminary figures 1990)**

Copper	36 458
Zinc	125 052
Nickel	57 812
Aluminium	845 068

Source: Central Bureau of Statistics of Norway

Inputs: electricty policy

Norwegian electricity production is almost entirely based on hydro-power. Statkraft SF, the state-owned power producer has about 30 per cent of the production. Statnett SF, the state-owned central high voltage grid company owns 80 per cent of the grid. In addition, there are 235 energy utilities mostly vertically integrated and mostly owned by local authorities. About 10 per cent of the power production capacity and some parts of the grid are privately owned.

The new Energy Act, which became effective on 1 January 1991, lays down the legal foundation for making the energy market more efficient. The organisation of power supplies is now based on a clear distinction between power production and sale which can be organised through a market, and the grid which is a natural monopoly. Norway has introduced common carrier to the grid.

Energy trade is a basic condition for a rationally functioning market. The Energy Act makes it possible for large consumers, like power-intensive industries, to buy from the cheapest supplier.

Some Norwegian industries own their own power plants. These are mostly old power plants with low costs. According to the concession, these power plants are reverted to the state without compensation after 60 years. Usually industry leases these plants following reversion.

Since 1947, the Norwegian power-intensive industries have signed long term state power contracts. The plants are base load consumers of large quantities receiving the power on high voltage. They are often located near a power-plant.

The state power contracts introduced in the 1950s, 1960s, 1976 and 1983 were based on covering the costs of the required investments in new hydro-power generating capacity. Prices have increased in the period because of increasing investment costs. The oldest contracts run until 2004 and 2011, while the contracts from 1976 run until the end of 1995.

From 1989 the State has offered 2.7 TWh/year for new industrial enterprises. For the power intensive industry, the 1991-price inclusive electricity tax and transmission charges, amount to 17.2 øre/kWh[1]. These are not firm power contracts. According to the terms of the contract, the contract holders have to pay the price of occasional power when the price exceeds a certain level. When the present surplus capacity decreases, and the price of occasional power rises, the actual price paid by the power intensive industry might therefore be higher on average than 17.2 øre/kWh. Being valid for a duration of 13 years, these contracts are not as long as the older ones.

In 1991 Parliament decided to offer to renew contracts expiring in the 1990s. At the same time the Parliament opened the way for Statkraft to re-negotiate the terms of all industry contracts in an ordinary business-like way. The industry's energy prices can, for instance, be linked to the product they sell. The renewal price being offered to the power-intensive industry is, on average, 17 øre/kWh in 1996, rising to 20 øre/kWh in 2010. This includes electricity tax and transmission charges. The price level refers to year 1991. The contracts run until 2010. In case the power market develops in a way that makes the terms of these contracts unreasonable, the contracts will be re-negotiated in 2005. Also for these contracts, Statkraft SF and the individual industry customer can agree to link the power price to the prices of the latter's products. However, 17 øre/kWh is a floor price.

In the present market these prices are relatively high. However, the prices are expected to increase. Due to this there will be a rise in the real prices for the firm price contracts. The level of the offered prices exceeds the average real historical costs connected to building and running the power plants.

The conditions offered to owners of power plants which are reverted to the state before 2000 are mainly the same as in the firm price contracts. However, these customers are not charged for transmission charges.

If all of industry renews the contracts on these conditions, it will make the average prices for all existing firm power contracts with the state power company rise from 11.7 øre/kWh in 1991 to 12.2 øre/kWh in 1996 and 13.3 øre/kWh in 2003. Electricity tax, which amounts to 4 øre/kWh, is included.

The latest available statistics indicate that the average price of electricity consumed by the power-intensive industry was 11.5 øre/kWh in 1989, electricity tax included. The price also includes self-generated hydro-electricity, power from local electric utilities, as well as occasional power. For all industry, the average price was 15.8 øre/kWh.

Government ownership

Norsk Hydro A/S — aluminium and magnesium (primary metal and processing)

The Norwegian State participates with direct ownership interests in Norsk Hydro A/S (51 per cent). Norsk Hydro, the parent company of Norway's largest industrial concern, is engaged in the following business areas: agriculture; oil and natural gas; petrochemicals; aluminium and magnesium; biomedicine; aquaculture. The company operates world-wide and its shares are quoted on stock exchanges in eight different countries.

In 1990 the company's production of primary aluminium metal, through the wholly-owned subsidiary, Hydro Aluminium A/S, amounted to 638 000 tonnes. The primary aluminium plants are all located in Norway. Approximately 60 per cent of the primary metal is processed at Hydro's own processing plants in western Europe and the US. Production of magnesium metal amounted to approximately 65 000 tonnes. The magnesium metal production plants are located in Canada and Norway and the metal is processed at subsidiary companies in Norway, Sweden and the UK.

The Norwegian State has held 51 per cent of the shares of Norsk Hydro since 1971. Its involvement in Norsk Hydro is limited to management of the ownership interest, including participation in share issues[2]. The company has not received any grants, loans or guarantees from the State.

A/S Bleikvassli Gruber

The Norwegian State participates indirectly in mining of lead and zinc through A/S Bleikvassli Gruber. The company is located in the Nordland county in northern Norway and is a wholly-owned subsidiary of A/S Sydvaranger (87.45 per cent owned by the state). Annual production volume at Bleikvassli is approximately 12 000 tonnes of zinc and concentrates and 5 000 tonnes of lead concentrates. Most of the production is exported to Germany under long term contracts.

Bleikvassli has for several years operated in the red and has been dependent upon state financial support through its owner A/S Sydvaranger. For 1989 and 1990, however, the company achieved profitability and was able to operate its business without external financial support. Remaining ore reserves are, at present production rates, estimated to last until approximately 1998.

Sulitjelma Bergverk A/S

Sulitjelma Bergverk A/S, a copper and zinc mine wholly-owned by the Norwegian State, closed down operations in June 1991. When necessary environmental and security measures have been finalised and mining equipment has been sold, the company will be wound up.

Regional policy incentives

Grants, loans and guarantees can be extended to companies located in the less-favoured regions of the country. There is no discrimination between sectors, and non-ferrous metal industries can apply for support in line with other sectors.

The Regional Development Fund (a state-owned bank) allocates the company specific support. The Regional Development Fund can support company or sector development both as loans and guarantees. Support for investment and company developments is for new establishments and innovation. Support to smaller or medium-sized companies is given priority. All projects are to be profitable even without the support.

The country is divided into four zones with differing maximum rates of investment subsidy, ranging from 40 down to 15 per cent. Every project is assessed in its totality and the support rate may be lower than the maximum rate in the zone. In addition to the four zones, there is a fifth zone for which no investment subsidy is given, but where the other support methods are operative.

Also, the country is divided into six zones with differing rates of employers' contributions to social security, which only partly correspond to the five zones mentioned above. For the central regions, the rate is 16.7 per cent, while there are no employers' contributions in the northernmost regions (Finnmark and Nord-Troms). This applies to both mining and metals industries.

Taxation

There are no special tax rules applying to the non-ferrous metals industry in 1991. General tax rules for industry are set out above in the section on mining taxation.

HORIZONTAL POLICIES

Environment

The Norwegian requirements for the regulation of emissions to the air and discharges into water, waste disposal, etc. are governed by the Environment Act of 1983. In Norway, it is the central government that administers the law, unlike other Nordic countries, such as Sweden and Denmark, for example, where there is considerably more delegation to local authorities and where the central government plays more of a guiding role. Every enterprise of a certain size in Norway obtains specific permits regarding the use of natural resources as well as pollution permits.

The emphasis of Norwegian environmental policy has changed from direct regulation to measures which serve to raise the cost of production of the polluter and/or the relative price of the polluting product. The 1991 National Budget calls for the evaluation of cost efficient taxes, both in cases where the actual emission of pollution can be accurately measured or where it is difficult to do so. The intention is to use the extra revenue from "green taxes" for lowering other taxes. In this context, the Parliament has decided on a series of unilateral measures. According to official estimates, the proposed carbon tax will reduce carbon-dioxide emissions from Norwegian manufacturing and transportation industry by 3 per cent at the end of this century. The high initial level of the carbon tax has, however, resulted in lower relative prices of imported electricity generated by coal, and concerns have been expressed that the tax as a consequence may contribute to increased emissions of carbon dioxide.

Mining

As mentioned above, mining of ores requires a mining concession, delivered by the Ministry of Industry. This concession holds conditions which, among other things, regulate the operation and closure of the mine. For any mining project to be authorised, it must first undergo an environmental and social impact study.

Norwegian environmental regulations are based on a discharge permit — issued on a case-by-case basis — and cover water pollutants, air pollutants, mine waste and tailings and management of toxic waste. A discharge permit also stipulates detailed regulations on production amounts and abatement equipment, etc. General effluent guidelines only exist for the electroplating industry. All ore mining requires a discharge permit issued by the State Pollution Central Authority.

Metals industry

For the nickel industry, limits are set for HCl, Cl_2, SO_2 and H_2S emissions. The limits are set in kg per product, with the exception of H_2S (limit: 8mg/m^3). The biggest environmental problems associated with nickel production are considered to be chlorinated or brominated organic compounds in waste water and the disposal of solid wastes. With regard to waste disposal (lead, cobalt, iron arsenates), the problem is solved at present by dispersal and depositing in mountain cavities.

The biggest environmental problems in the aluminium industry are fluoride and poly-aromatic hydrocarbon (PAH) emissions. Limits are set for fluorides (10-20kg/hour), PAH, particulates and SO_2.

As regards acid rain, Norway has entered into international agreements where the aim is a 30 per cent reduction in SO_2 emissions between 1980 and 1993. Norway has committed itself to a 50 per cent reduction by this date. All aluminium smelters have installed desulphurisation plants (seawater scrubbers). The 50 per cent target was achieved in 1990.

In 1988-89 the Norwegian enterprises in the state sector have prepared, on behalf a request from the government, a synopsis of the environmental aspects of their activity (scale of the environmental measures, time-scales for implementation, and costs entailed by the measures) and measures to cope with. The planning and implementation of concrete measures depend on negotiations with governmental agencies.

Several firms submitted their own reports, like Norsk Hydro, Norsk Jern Holding A/S and others. They were focused on safety and increasing environmental control through modernisation of the plants.

With regard to the production of aluminium, Norsk Hydro recognizes that there is some potential for reducing the discharges of $C0_2$ in the production of primary aluminium. Currently, the discharge of $C0_2$ per tonne of produced primary aluminium is higher in the case of Söderberg cells than in case of prebaking cells. The overall reduction potential is almost 5 per cent. This is important also in the context of the recently adopted carbon tax which exempted the aluminium industry (as it did certain other industries) on its carbon-specific emissions and thus will have no direct influence on reducing the above mentioned CO_2 discharges.

Health and safety

Labour Inspection controls heath and safety regulations. These regulations apply to the whole range of working activities and involve for example the quality of working environment, exposure to noise, vibration, dust and fumes.

Trade policy

Norway is a low tariff country. Duties on imports of non-ferrous ores/ concentrates and unwrought metal are zero. The simple unweighted average of tariffs levied on semi-manufactured products of non-ferrous metals is 1.4 per cent, or calculated on an applied basis, 0.1 per cent. Certain aluminium semis face a 9 per cent tariff for MFN countries, but most trade enters on a preferential basis.

Under the regulations with respect to the licensing of strategic goods, restrictions are applied to mitigate shortages of raw materaials for the Norwegian metallurgic industry. EFTA countries and the EEC signed bilateral agreements concerning the elimination of existing restrictions on non-ferrous scrap, which entered into force in January 1990.

Following anti-dumping investigations instigated by the EC Commission on Norwegian exports of ferro-alloys to the Community, the Norwegian industry has agreed to price undertakings involving respect of minimum prices.

Research and development (R&D)

The Royal Council for Scientific and Industrial Research (NTNF) supports applied research and development, programmes with financing of up to 50 per cent of a project. Priority areas are information technology, materials technology, biotechnology, environmental technology, aquaculture and offshore technology.

Within the priority area of materials technology, NTNF can support R&D, including on environmental technology, that aims to improve the production processes of materials, and R&D for developing further utilisation of materials produced in Norway. A major proportion of this support has been given to projects that aim to develop aluminium and magnesium alloys and new applications and products from these light metals.

Foreign investment controls

As described above, foreign citizens have to establish a company with its seat in Norway in order to obtain a concession, issued by the Royal Ministry of Industry. All applications from foreign-controlled companies made in the 1980s were approved by the Ministry of Industry. Currently the law concerning such foreign investment is undergoing revision.

CONSUMPTION/SUPPLY

Securing overseas supply

Norway has no particular programmes to assist Norwegian producers abroad in exploring for mineral resources, nor in support of mining projects generally.

Stockpiles

No government stockpile programme is in operation.

Recycling

Concerning problems of waste, there is a general consensus that the effort to achieve the measures of sustainable development has to be based on the following main strategy:

- minimising waste;
- encouraging recycling;
- treating remaining waste in an environmentally acceptable manner.

The government has introduced a programme for cleaner technology to promote waste minimisation. The financial contribution to this programme was NKr 54 million in 1991. The measures of the programme are to introduce and test different methods to make changes in product design and production processes in order to reduce the total amount of waste and to economise available natural resources, including raw materials.

It is also the government's objective to reduce the amount of waste by recycling waste as far as possible. Both economic and other means will be used to stimulate sorting of waste and to encourage private enterprises to increase the use of refuse-derived raw materials and refuse-derived fuel. The government proposed to increase the financial contribution for recycling up to NKr 55 million in 1992. Neither this nor the above-mentioned programme for cleaner technology are specially oriented towards the non-ferrous sector. Norway has a small number of secondary aluminium plants. The order of magnitude of the output of these plants is nevertheless at present minimal as compared with the primary aluminium industry.

Laws and regulations provide opportunities to require source reduction and recycling. Both the Pollution Control Act and the Product Control Act have an important impact on enterprises and municipalities as well as the whole society. The State Pollution Control Authority can instruct enterprises to sort and deliver waste to another enterprise. On the other hand, dealers can be made responsible for taking back products after they have turned to waste.

When imposing requirements regarding the treatment of waste, costs will be raised. This will also make projects based on source reduction and recycling more profitable and society will gain economically from reduced quantities of waste. Another effect is to increase the amount of raw materials that can be recycled relieving pressure on limited natural resources.

NOTES AND REFERENCES

1. 1 øre = 1.5 thousandths of US$
2. The company was established in 1905 with predominantly French capital to manufacture the world's first industrially produced nitrogen fertiliser. Diversification dates back to the 1950s.

BIBLIOGRAPHY

Central Bureau of Statistics of Norway (1990), *Natural Resources and the Environment 1989*, Oslo.

GATT (1991), Trade Policy Review Mechanism, Norway.

KIRCHNER, Christian (1988), "Western Europe: Subsidized Survival", *The World Aluminium Industry in a Changing Energy Era,* in Peck, M.J. (ed.), Resources for the Future, Washington D.C.

Mining Magazine (1987), "Norwegian Smelters", London.

Norsk Metallurgisk Industri (1990), *Perspektiver 2000.*

OECD (1991), *Recent Trends in the Regional Situation and Policies*, Norway.

Royal Ministry of Industry (1988-89), *Bedrifter hvor Staten* v/Nærings-departementet har eierinteresser.

SVENDSEN, Bjørn (1989), *Norsk aluminiumsindustri — Situasjon og utsikter,* Bergen.

SVENDSEN, Bjørn (1991), *Virkninger av en internasjonal klimaavatale på norsk aluminiumindustris konkurranseevne og fremtidsutsikter*, Bergen.

PORTUGAL

INTRODUCTION

Portugal's mining and metallurgical industry was known for its production of ferro-alloys and tungsten, whereas non-ferrous metals production was modest. In the 1980s this pattern changed with reductions in ferro-alloy and tungsten production and increase production of non-ferrous metals. Between 1988 and 1990 mineral production rose by nearly 134 per cent by value and exports by 184 per cent, to account for 3.4 per cent of Portugal's total exports.

This was largely due to the coming on stream of the mine at Neves Corvo, a complex multi-metal deposit, yielding copper (with zinc and silver as by-products) and tin. Reserves are an estimated 32 million tonnes of copper at a metal content of 8 per cent. Present production capacity is about 115 000 tonnes per year (tpy). Since April 1990 Neves Corvo has been producing tin. Neves Corvo is controlled by the Sociedade Mineira de Neves Corvo (Somincor), a partnership between the state firm Empresa de Desenvolvimento Mineiro (EDM), which owns 51 per cent of Somincor, and Rio Tinto Zinc (RTZ), which owns 49 per cent. The European Investment Bank (EIB) has funded 50 per cent of the project's development costs.

In 1990, Couto Mineiro de Aljustrel reduced its production of cupriferous pyrites due to reduced capacity at the sulphuric acid plant. As a result the company installed a complex ore processing plant with a nominal annual capacity of 6 300 tonnes of copper, 27 500 tonnes of zinc and 4 400 tonnes of lead. The state holding company EDM has majority control of Couto Mineiro de Aljustrel in partnership with the Swedish company Trelleborg.

Portugal is the only EC country still producing tungsten concentrates at the Couto Mineiro de Panasqueira mine (1 430 tonnes of tungsten content in 1990), which is owned by Beralt Tin & Wolfram, Portugal SA. This firm is investing in improving productivity with assistance from regional development programmes. Recourse to anti-dumping provisions resulted in the application of anti-dumping rates on tungsten concentrates from China for a limited period. Also the EC Council of Ministers approved a Resolution (No. 74/86) introducing a financial support scheme for suspended tungsten and tin workings to safeguard mining infrastructures. The scheme's main activity has been to study the economic and commercial feasibility of re-opening mines at Montesinho, Argozelo and Vale das Gatas.

Portugal's membership in the EC has stimulated the interest of leading mining and metallurgy companies for prospecting and exploring in Portugal. Its underground is relatively unknown from the geological point of view and geologists suspect that it bears large unexploited reserves of non-ferrous metals.

PROSPECTING AND EXPLORATION

The Geological Resources Act of 1990 replaced legislation dating back to 1928. The Decree 90/90 applies to all mineral resources and it distinguishes public resources from resources under the private law regime. Prospecting, exploration and mining activities of public and private resources are governed by specific legislation. The private law regime is confined to certain non-metalic resources (spring waters and quarries) while mineral water and geothermal resources and ore deposits belong to the state.

Geological services

The Directorate General for Geology and Mines (DGGM) is responsible for surveying and exploring Portugal's geological resources and to carry out their exploitation. Its main responsibilities are to contribute to the defining of state mining policy, to conduct geological and hydrogeological surveys and inventories of Portugal's geological resources, to provide technological supports to the industry, and to co-operate with the relevant regional development and environmental agencies.

The DGGM's main strategy can be summarised as follows:

a) developing geological and hydrogeological mapping, especially of lesser-known areas;

b) undertaking a comprehensive inventory of mineral resources and carrying out their exploitation by state or private firms;

c) improving the characterisation, processing and production of minerals, particularly by modernising plant technology;

d) defining mining policies and improving the production and use of geological resources.

The DGGM is also in charge of three services, the Portuguese Geological Service, the Mining Development Service (SFM) and a Laboratory.

a) the Portuguese Geological Service (SGP) is responsible for processing and disseminating geological information and for the preparation of geological maps of the country. Services available to the public include advice on scientific and technical matters concerning geology and mineral resources, and information and assistance in various fields of geology;

b) the SFM identifies mineral resources and is therefore the state's main instrument for the development of mining activity in Portugal. It makes a systematic inventory of mineral resources, plans and implements prospecting programmes, conducts R&D in new prospecting techniques, develops new methods of processing geological information, and circulates this to potential users. The SFM provides services to the public in prospecting and exploration, and the identification and evaluation of mineral resources;

c) the laboratory is a centre for applied research and it carries out studies on minerals characteristics, pilot stage ore processing and environmental controls.

Affiliated to the DGGM, the Advice, Promotion, Administration and Co-ordinating Service for the Extractive Industry is responsible for negotiating and awarding contracts for prospecting, exploring and exploiting resources of the public domain resources and for advising on mining policies and fiscal and financial investment subsidy awards.

Exploration permits and concessions

Portuguese law, with its tradition of regalian rights, distinguishes ownership of the land from ownership of the underground. Underground resources belong to the state which exercises a right of ownership in respect of immovable public or administrative property. This principle is enshrined in the Constitutions of 1933 and 1976.

Prospecting, mineral exploration and exploitation of public domain resources require corresponding permits from the Secretariat of State for Industry, under the Ministry for Industry and Energy. Exploration permits may be issued upon request or auctioned. The DGGM is also empowered to award licences for large-scale mineral exploitation and to monitor compliance with relevant legislation and regulations. Local authorities are empowered to award small-scale operating licences, especially for quarries.

Taxation

Mineral prospection and exploration costs are considered as operational costs for the financial year in which they were incurred and do not qualify for special tax treatment.

Direct participation

The state participates directly in exploration activities through the Mining Development Division (DFM). Exploration work is either performed directly by the DFM or it is contracted to specialised firms. Deposits identified by government services are awarded as concessions to private firms, upon application or by auction, under conditions which vary according to the quality of the deposit, the amount of public expenditure incurred, the value of the ore and the possibilities of exploiting on a commercial basis.

The EC has provided financial assistance for exploration-related studies. The first project involves the development of prospecting techniques for gold and precious metals in the western part of the Iberian Peninsula. The second project involves multidisciplinary research into gold-bearing lodes in the mountains of western Spain and Portugal. These precious metals projects are being implemented jointly by the DGGM in co-operation with other institutions, including the Spanish Oceanography Institute, the Woodshole Oceanographic Institution and the Institute of Oceanographic Sciences.

MINING AND METALS PRODUCTION

Mining

Minerals taxation

The corporation tax is set at 36.5 per cent in Portugal. In addition companies have to pay a range of central and local taxes. Projects regarded as being of strategic importance may qualify for tax concessions relating to the depreciation of investments in equipment, these concessions have to be negotiated with the state. The law also permits the application of higher depreciation allowances when industrial assets are used intensively and for exceptional events (unforeseen falls in value, rapid technological innovation). Industries regarded as having priority for Portugal's economic development may qualify for a 4 per cent reduction in the state levy on real estate acquisitions.

Very large-scale industrial projects, involving an investment in excess of Esc 10 billion, which will contribute significantly to exports, may qualify for tax concessions, i.e. total or partial exemption from corporation tax, from the levy on real estate acquisitions and from local taxes. These tax concessions must be negotiated with the state.

Metals production

The non-ferrous metals sector is listed as one of the priorities of the government's industrial policy. The state has established mechanisms to speed up the industrial development through productivity, quality, the environment, innovation, energy saving and training programmes. The government has thus decided to develop Portugal's mineral resources by maximising the domestic processing of national resources. In this context, the government has been studying the economic feasibility of a copper smelter/refinery, a project in which the Finnish firm Outokumpu Oy has shown interest.

Another sectoral priority in the government's industrial policy is the modernisation and restructuring of mature industrial sectors. In this context the Programme for the Restructuring of the Foundry Sector was set up, under which financial assistance is available to privately operated foundries wishing to modernise their productive capacity and to improve their competitiveness.

Inputs

Portugal's scarce energy resources and the long distance of these resources from consumption centres mean its energy prices are among the highest in Europe. There is no programme favouring the heavy energy consumers. Similarly, there is no preferential tariff policy with regard to transport.

Direct and indirect support

Mining firms may qualify for direct or indirect state aid through technical work, subsidies and tax concessions. Since 1988, Portuguese mining and metallurgical firms are also eligible for assistance under special EC schemes as part of the structural adjustment programmes for the integration of Portugal's economy within the EC. These include the Specific Industrial Development Programme for Portugal (PEDIP) and the Regional Incentives System (SIBR).

The PEDIP aims to modernise Portugal's industry and to make it more competitive, thereby preparing it for integration within the single European market. Under the PEDIP, financial incentives are available for projects in the least industrialised regions, to strengthen their production structures and technological base, to stimulate innovation and to improve environmental protection. The PEDIP offers Portuguese enterprises a range of investment grants, particularly for the building of infrastructures, staff training, and productivity improvement.

To support regional development, the Regional Incentives System (SIBR) programme was established in 1989. It is funded at 70 per cent by the European Regional Development Fund (ERDF). Assistance under this programme takes the form of direct participation in the funding of projects. The government recently empowered the PEDIP to fund projects for rationalising use of energy, for improving productivity and for manpower training in the non-ferrous metals processing sector.

Government ownership

Since the "Carnation Revolution" of 1974, the state's involvement in the economy has increased considerably through nationalisation of several mining and metallurgical units. The Empresa de Desenvolvimento Mineiro (EDM) was accordingly established to consolidate and manage the state's interests in these various mining activities, while state involvement in the metallurgy sector has been mainly through the State Participations Institute (IPE), a government holding company. The statutes of state enterprises have been changed recently with a view to privatising part of their capital. However, the state will be retaining a majority interest only in enterprises of a strategic character.

HORIZONTAL POLICIES

Environment

The Environmental Assessment Act (DL 186/90) sets limits on the discharge of pollutants into the atmosphere. Portugal also has a General Regulation (DL 162/90) on industrial safety and health in mines. The Environmental Assessment Act provides that any project which is liable, because of its nature, scale or location, to affect the environment seriously must first be subject to an environmental impact study. DL 352/90 establishes limits and reference values for air pollution concentrations and limits for emissions from industrial plants. Special provisions covering mineral exploration and exploitation call for the preservation of environmental quality and restoration of the site. If annual production is to exceed 150 000 tonnes, no mining permit will be issued until an environmental impact study has been done.

For extractive operations that are governed by private law, particularly quarrying, a favourable assessment with regard to the environmental aspects must first be obtained from the Regional Co-ordinating Committees (CCR) and from the National Parks, Reserves and Nature Conservancy Service (CNPRCN) before an operating permit is issued. Just as for mining, if annual production is to exceed 150 000 tonnes, no quarrying permit will be issued until an environmental impact study has been done.

Under this legislation, the DGGM's activities are extended to environmental conditions in mines, in particular monitoring levels of dust, noise, gas, temperature and vibration. Portugal is following very closely the work that is being done by the EC on health and safety in mines.

Research and development

The DGGM's technical laboratory helps Portuguese producers to identify new technologies and improve existing ones. The laboratory's work is concerned with the characterisation of basic material, experimental and pilot-scale processing of industrial ores, chemical analysis, environmental management in mines and quarries, applied research projects in metallurgical engineering and prospecting for mineral resources.

Trade policy

As a Member of the EEC, Portugal applies the common external customs tariff, the section of which relating to the six main non-ferrous metals is summarised in the chapter dealing with the European Economic Community. Portugal's accession agreement to the EEC provides for certain quantitative restrictions on semi-finished copper products; these expire at the end of the transitional period.

CONSUMPTION/SUPPLY

Securing overseas supply

There is no specific government measure to stimulate or financially support the development of mineral prospecting or exploitation abroad.

Stockpiles

Portugal has never maintained any strategic stockpiles of metals.

Recycling

The Portuguese Government is anxious to develop recycling because of its economic and environmental importance. The Directorate General for Industry (DGI) has promoted the establishment of a representative association for recycling and has removed obstacles to treating recycling projects for PEDIP fund purposes. The DGI has also been working to interest private firms in participating in R&D activities financed under the BRITE-EURAM programme (raw materials and recycling). The DGI is carrying out, with the EC Commission, a feasibility study on a secondary aluminium project.

BIBLIOGRAPHY

Boletim de Minas, Vol. 26, n° 2, Lisbon, 1989.

Direcçãó-Geral da Indústria: *Estratégia de desenvolvimento de metais não ferrosos.*

Ministério da Indústria e Energia, *Direcção-Geral de Geologia e Minas: Relatório de Atividades*, Lisbon, various years.

PEREIRA, A. R., *Aspectos Gerais do Desenvolvimento do Sector Extractivo.*

Power in Europe, 50, 25 May 1989.

SPAIN

INTRODUCTION

Spain's position in mining for three of the six main non-ferrous metals is not inconsiderable. At European level, it is the leading producer of zinc ore and lead. In copper mine production, although currently in decline, Spain still ranks first in the EEC. And in the production of copper blister, it ranks second immediately after Germany. It has a large non-ferrous metals processing industry and is heavily geared to exports in the field of zinc and aluminium.

More abundant and less extensively exploited than in most of its European neighbours, Spain's mineral resources have in the past attracted investment from mining companies abroad, in particular from countries now Members of the EC, for example Rio Tinto Zinc (RTZ) and Peñarroya.

The development of Spanish industry has long taken place within the context of a closed economy, protected by import licences, quota-systems, high tariffs and a deterrent taxation system. Starting in the 1960s, Spanish industry gradually began to specialise in a certain number of basic sectors, such as non-ferrous metals mining and processing, iron and steel, motor vehicle construction, shipbuilding and chemicals.

In this context, the National Plan for Mines of 1971 and the Mining Law of 1973 sought to encourage the development of the Spanish mining and metallurgical industry under stricter control by the national government. There are a large number of provisions for government intervention in the development of the mining and metallurgical industry, in particular the regulation of foreign investment and the possibility of setting aside areas of the national territory in relation to their importance for economic development or national defence.

These provisions constitute the basis of Spanish mining legislation, while the taxation system applicable to the mining sector is partially defined in the Law for the Encouragement and Development of Mining of 1977. Pursuant to this Law, a National Raw Materials Supply Plan (PNA) was approved in 1978. It defines objectives and sectoral programmes for the exploration and exploitation of those mineral resources accorded priority and hence of national importance. The main objectives of the PNA are to reduce dependence on and vulnerability to supplies from abroad and to promote the development of mining and metallurgical activities on Spanish territory. The sectoral objectives and programmes laid down in the Plan are updated every two years.

The difficulties of many industrial sectors prompted the Spanish Government in the 1980s to implement a number of plans to assist the restructuring of enterprises in difficulty in the various sectors including aluminium, copper processing and steel. The main measures involved financial support for initiatives designed to reduce costs, to adjust production capacities to trends in demand and to promote measures to restore the financial health of companies.

. The general aim of government intervention in the mining sector is to maximise the exploitation of mineral resources by way of planning at national and regional level. The Ministry for Industry, which is responsible for the application of the Mining Law, regulates the private sector and, through the intermediary of the Instituto Nacional de Industria (INI), a State holding corporation set up in 1941, directs most of the state enterprises. The Ministry also prepares the National Supply Plan which determines priorities and specifies the main lines of government intervention in the mining and metallurgical industry.

Finally it should be pointed out that, under the Constitution of 1978, the national territory was divided into 17 Autonomous Communities. This political and administrative decentralisation resulted in the decentralisation in many fields of activity, in particular mining legislation and environment.

PROSPECTING AND EXPLORATION

Geological services

The Ministry for Industry has set up a Register of Mines which centralises all information available on the current status of mining rights anywhere in the country. Geological services are provided by the Instituto Tecnológico Geominero de España (ITGE), which is responsible for the preparation of the country's geological maps. Statistics are collected by the Statistics and Information Service of the Ministry for Industry and Energy.

Lands availability

The Act 4/1989 on Protected Natural Sites aims at protecting natural resources while allowing their exploitation in a context of an orderly economic and social development. Prospecting, exploration and exploitation activities are allowed in protected natural sites such as national parcs and reserves, natural monuments and protected landscape, if a declaration is submitted which provide for compatible objectives of these activities with those of the natural sites. This compability declaration has varying degrees of requirements, the highest being applied to parks and the lowest to landscapes.

Prospecting and exploration permits

Spanish law states that deposits of natural origin and other geological resources occurring on the Spanish territory, territorial waters and the continental shelf fall within the public domain. From the point of view of exploration and extraction, the law divides geological resources into three different groups: (i) "Group A" includes those which have little commercial value and whose marketability is geographically limited; (ii) "Group B" includes mineral and thermal waters; (iii) "Group C" comprises other mineral resources, hence non-ferrous minerals. In 1980 a new group, "Group D", was added to include carbon-bearing minerals, radioactive minerals, geothermal resources and other natural resources of value from the point of view of energy.

Responsibility for issuing prospection, exploration and exploitation permits has been transferred to the Autonomous Communities. If two or more Communities are affected, the competent authority is the General Directorate for Mines and Construction. At present, this Directorate has these responsibilities in Castilla, La Mancha, Cantabria, La Rioja and the Balearic Islands.

Prospection permits are valid for a maximum of two years and prospection works must not lead to substantial alteration of the land. Exploration permits are granted for three years and may be extended for another three years if results are promising, and extended thereafter at the discretion of the competent authority. Mining concessions are granted for periods of 30 years, renewable at 30-year intervals up to a maximum of 90 years. Direct or indirect acquisitions of mining rights by a foreign government require an authorisation from the Spanish Government.

Tax incentives

Government credits and capital subsidies of up to 20 per cent of the investment committed may be granted to Spanish enterprises for their exploration and mining operations. The conditions and forms of these subsidies or credits are defined on a case-by-case basis when the plan of work is drawn up. In the event of a commercially viable discovery, subsidies received for exploration must be repaid. Repayment is due within a period of 5 to 20 years after the start up of mining operations in accordance with the conditions agreed upon when the plan of work was approved.

Direct participation

The state may set aside areas of the national territory (in the wider sense of the term) in terms of their specific importance to Spain's economic and social development and to national defence. These reserves take three forms:

- Special reserves: activities relating to one or more stated resources are reserved for the state over the entire territory. They are set aside for a maximum of five years. This period can only be extended by legislation.
- Provisional reserves: these reserves relate to all or a few specific mineral resources in specified zones. They are set aside for the period laid down in the regulation, though the periods of one year for

prospection, Three years for exploration and 30 years for extraction may not be exceeded.

— Definitive reserves: these are created for the exploitation of resources located in clearly defined zones within provisional reserves. They are set aside for a term equivalent to that of provisional reserves. For a provisional reserve to be declared a definitive reserve, a commercial find must be proven.

Measures to set up reserves are taken by the Ministry for Industry. Such measures establish state priority over the available lands but must not prejudice mining rights already in existence. Within these reserves, prospection, exploration or mining activities may be pursued directly by the State, by partnerships between the state and Spanish and/or foreign private firms or by private operators tendering for public contracts.

Thus, for example, Decree No. 1018 of 1984 set aside for the state a provisional reserve for the exploration of lead, zinc, silver, antimony, copper and gold deposits in the Rio Gibranzos area in the Province of Caceres. The exploration work was performed by ITGE. Three years later, in the light of the results obtained, the Ministry for Industry decided that the investigations should continue. Decree No. 270 of 1987 transferred the reserve to INI and the work was continued by Empresa Nacional Adero de Investigaciones Mineras SA (ENADIMSA).

The prerogative of state intervention in the mining industry is clearly reflected in the provision stipulating that the state can directly undertake exploration and exploitation of deposits, particularly within areas of reserves set aside for it. The state thus has the right to intervene directly by the control or transfer of mining rights in order to maintain a presence in all phases of a mining project. Public or private firms taking on activities of the state naturally qualify for subsidies and subsidised interest rates to finance their operations.

Restrictions are applied on minerals deemed strategic by the Council of Ministers. Until now, no mineral has been declared strategic. As regards strategic minerals, mining rights for such minerals would be granted only to Spanish nationals or firms whose capital basis is entirely controlled by a Spanish interest. In the event that some minerals are declared strategic while their exploitation is underway, the government can request the holders of mining rights to supply the state with sufficient quantities to maintain a strategic stockpile.

MINING AND METALS PRODUCTION

Mining

General regulations

In the case of state-owned land, it is customary though not mandatory for operations to be carried out by ITGE and ENADIMSA, but this does not prevent other operators from being involved as well. The policing of mines is assumed by a service concerned with mining safety and it reports to the corps of mining engineers employed by the Ministry for Industry and Energy. In cases of emergency in which the safety of workers is endangered or the mineral resources, the installations or the environment are at risk, the relevant Autonomous Community may suspend operations.

In addition, the state has a legal obligation to support the setting up of operating consortia or syndicates (*cotos mineros*) by granting incentives and fiscal advantages. The operating consortium or syndicate is an interest grouping of holders of mining rights in different areas of one or more deposits located in such a way as to permit the joint use of all or part of the services necessary for their exploitation. These syndicates may be voluntary, or compulsory in the case of strategic materials required for national defence. In the same context, the law gives the government authority to designate particular mining sectors, or parts of them, as "industries of preferential interest" and to declare certain mining regions as favoured zones for industrial development, one example of such zones being Santiago de Compostela in Galicia.

Taxation

The holders of mining rights pay an annual duty assessed according to area. Spanish mining companies are also subject to the general taxation system (the tax rate applied is 40 per cent). However the Law for the Promotion of Mining of 1977 defined a certain number of promotional measures for which mining companies extracting ores designated as priority minerals are eligible. Thus, for example, in respect of their activities in connection with minerals designated as having priority under the PNA, operators qualify for a reduction of up to 95 per cent in the following taxes:

- general property transfer taxes and stamp duties;
- customs duties on imported products which are not produced in Spain;
- general tax on interest earned on capital and on loans obtained from international organisations.

These fiscal advantages are granted for five years and may be renewed for the same period if social and financial conditions allow. An exceptional five-year extension may be granted in the event of positive results.

Mineral raw materials initially identified as priority resources in which Spain has a shortage are: aluminium, copper, chromium, tin, manganese, iron ore, nickel, lead, titanium, tungsten, zinc, asbestos, carbon minerals, phosphates, uranium and other geothermal resources. These priorities are updated every two years.

An ore depletion reserve can be set up to a maximum of 30 per cent of the assessment basis for corporate tax. However the Government may raise this limit. If a priority raw material is involved, the operator may opt to deduct 15 per cent of the value of the ores sold. The sums thus deducted must be invested within 10 years in exploration work or in holdings in exploration companies, or they must be reinvested in mining operations.

Since 1977, expenditures for exploration and exploitation activities are depreciated for tax purposes on a 10 year basis. Subsidies are also available for Spanish companies equivalent to 20 per cent of investment committed or costs incurred in the transport of mineral raw materials.

Metals production

Inputs

Energy prices and rail freight rates (charged by the Spanish State Railways, RENFE) in Spain are relatively high compared with other European countries, which is often cited by the country's mining and metallurgical concerns as a major handicap to their international competitiveness. There is no special rate for freight transportation of minerals and none is expected. However, there are special rates for large electricity consumers such as aluminium and zinc producers.

Direct support

Direct support to the mining industry is defined in Act No. 6 of January 1987 on the Development of the Mining Industry and in the Ministerial Order of October 1987 which regulates the granting of aid in pursuance of the new mining policies. Aid will be granted to projects for geological and mining exploration, technological research and development and the mining and processing of minerals and projects relating to the environment. The Spanish State also allows mining and metallurgical enterprises to borrow money from banks offering as security the mining rights of deposits for which the feasibility of exploitation has been economically demonstrated.

Indirect support

The industrial redevelopment policy was launched in 1984 and applies to sectors particularly hard hit by the economic crisis, such as aluminium, copper processing, iron and steel and chemicals. Under this policy, the State has granted aid in the form of subsidies and loans at subsidised interest rates from the Banco de Crédito Industrial and sometimes in the form of participatory credits. Between 1983 and 1988, Ptas 1 350 billion has been allocated to carry out this programme.

In particular INI has also contributed to the financial and industrial restructuring of the Spanish aluminium producers. The support provided between 1982 and 1984 led to the creation of the company Aluminio Español, which is controlled by the majority shareholder Inespal, itself a subsidiary of INI. The industry fell completely under State control at the start of 1990. INI is also the owner of all equity shares of the Minas de Almagrera SA. INI has no participation in any other non-ferrous metals firms. It participates into various regional development and industrial promotion organisations. Fifty-one per cent of

Rio Tinto Minera belongs to the Spanish Group ERCROS, with the remaining 49 per cent being held by the British company RTZ. Rio Tinto Minera has never been nationalised.

Within the regional industrial restructuring programme put into effect in 1984, a wide-ranging programme was established to reduce capacities in industries in decline and to modernise those which were economically viable. So-called "ZURs" (Zones for Urgent Reindustrialisation) were created with the provision of fiscal advantages, in particular, equipment credits at preferential rates and subsidies to encourage the establishment of new production units. Moreover, extensive new financial aids were extended to persons losing their jobs as a result of restructuring. In all regions, the mining industry and mining resources are considered suitable activities for development. On 31st March 1989, the Community Support Fund for regions defined as having top priority was set up, with a budget of 9 779 million Ecus. The regions concerned are Andalucia, Asturias, Castilla-La-Mancha, Castilla-Léon, the Canaries, Estremadura, Galicia, Murcia, the Community of Valencia, Ceuta and Melilla.

Finally, in 1990 the European Commission adopted the Community framework of support for Spanish regions in industrial decline. Seven regions will benefit from Community support to a total value of ECU 650 million from the Social Fund and the European Regional Development Fund (ERDF), namely: Catalonia, Basque Country, Madrid, Aragon, Cantabria, Navarra and La Rioja.

Government ownership

The most important instances of State intervention in the mining and metallurgical industry arose following the crisis of the 1980s. They mainly affected the aluminium industry — Aluminio Español SA, controlled by the holding company Inespal — which came under State control in 1984 after the withdrawal of Pechiney. Moreover, the minority private shareholder, the Canadian company Alcan, sold its shares to Inespal in 1990. In addition, the copper processing industry in Spain is operated by the company Rio Tinto Minera (RTM), a joint undertaking consisting of RTZ (49 per cent) and the State enterprise Unión Explosivos Rio Tinto (51 per cent).

Tax incentives

Mining and metallurgical enterprises are allowed to depreciate their investments in mining assets over a period of 10 years. Moreover the extraction of minerals accorded priority under the PNA may be exempted for 5 years (renewable for a fresh term of 5 years) for up to 100 per cent of a certain number of taxes and duties, in particular the tax on capital income and the tax on property transactions.

HORIZONTAL POLICIES

Foreign investment controls

Under Decree 1303/86, mining concessions may be owned by natural or legal persons of Spanish or foreign nationality. Participation in Spanish companies by the International Finance Corporation is not subject to this autorisation. In the case of mineral raw materials of strategic interest, mining concessions may be granted only to natural persons of Spanish nationality or to companies whose capital is entirely Spanish. It is for the Council of Ministers to declare raw materials as being of strategic interest.

Environment

The Decrees 1302/86 and 1131/88 on environmental impact assessment, require submission of an environmental impact assessment statement before the granting of mining concessions. The Decrees also specify the criteria retained for completing the assessment. These Decrees may be strengthened by the relevant Autonomous Communities competent in respect of mining matters.

Government credits and capital subsidies may be granted to Spanish companies for their environmental protection activities when these are required by the administration. The conditions and forms of such credits or subsidies are defined on a case-by-case basis when plans of work are drawn up.

Standards for the control of non-ferrous metal emissions to the atmosphere fall within the jurisdiction of the Ministry for Industry and Energy and are laid down by Decree 833/75 dealing with the protection of the atmospheric environment. For lead dust emissions, the limits are laid down individually for each plant, depending on its size and age.

For plants emissions from which are less than 300 mg/m^3, the limits are 120, 100 and 80 mg/m^3, respectively, for those built before 1975, between 1975 and 1980 and after 1980. For plants emissions from which are above this level, the limits for plants in the three age categories are 20, 15 and 10 mg/m^3, respectively. Emissions at ground level are also limited to 50 $\mu g/m^3$ (for a maximum of 30 minutes) or 10 $\mu g/m^3$ (for a maximum of 8 hours). The maximum quantity of lead permitted in the air of communities in the vicinity of metallurgical plants was fixed at 2 mg/m^3 in 1987.

With regard to the control of lead in industrial effluent, the limits in force since 1977, fixed by the Ministry of Public Works, require that 50 per cent of discharges made directly into the sea should be less than 7.5 mg/litre, while the limit of 15 mg/litre may not be exceeded by over 10 per cent of discharges. Decree 849/86, lays down limits equal to 0.5 mg lead per litre for discharges to rivers and lakes. Limits on lead at the work place in Spain were fixed in 1961 at a maximum of 0.15 mg/m^3 air. The maximum level for lead in the blood of workers was fixed at 80 $\mu g/100$ ml in 1963.

Research and development (R&D)

One of chief handicaps of Spanish industry is undoubtedly its late technological advancement. Indeed, in order to catch up, Spanish firms have concentrated on the acquisition of patents and licences from foreign companies rather than investing in research and development. As far as the mining and metallurgical industry is concerned, there have been no special measures in the past to boost R&D investment. In 1978 ENADIMSA was given responsibility for the execution of studies and projects called for by the PNA and concerned mainly with basic scientific research on ores and with metallurgical processes for the treatment of complex ores and the upgrading of by-products.

Nevertheless, this sector was strengthened by a Decree of 1986 which authorised the granting of subsidies for the financing of research and development activities. In this context, a new "Plan Nacional de Investigación Científica y Técnica" was prepared for the period 1988-90, providing for subsidies from the Ministry for Industry and Energy for technological research programmes of enterprises and trade associations. Sectors such as metallurgy, metallurgical products, electrical equipment, materials and the environment have received financial aid for R&D projects.

Moreover, the Centre for Technological Development in Industry (CDTI) provides financial aid for technological development projects in various sectors, including new metallic materials. CDTI also provides the management for international programmes, such as EUREKA.

Trade policy

As a Member of the EEC, Spain applies the EEC's common external customs tariff, the section of which relating to the six main non-ferrous metals is summarised in the chapter dealing with the European Economic Community. In the agreement on the accession of Spain to the EEC, certain quantitative restrictions were negotiated for copper tubes and pipes; these will expire after the transitional period for accession.

CONSUMPTION/SUPPLY

Securing overseas supply

In the context of the PNA, mining exploration programmes have been developed within the framework of bilateral co-operation agreements with the governments of Ecuador, Peru, Mexico and Equatorial Guinea. At the same time, ITGE has concluded several scientific and technical co-operation agreements with the national institutions of various countries in Europe and Latin America, on the lines of the arrangement between the French Bureau de Recherches Géologiques et Minières (BRGM) and the Government of Colombia.

Stockpiles

Spain's strategic stockpile is limited to uranium, in the form of yellow cake. Law 6/77 provided for the creation of commercial stockpiles of other mineral raw materials in the event that that should prove necessary for the stability of supplies to Spanish industry, as far as assets considered to be of national importance are concerned. However this facility, which in particular allowed for the participation of private firms in the financing of the stockpile, has never been put into effect.

Recycling

There is currently no programme to stimulate the recycling of non-ferrous metals.

BIBLIOGRAPHY

Gaceta de Madrid (1973), Law 22 of 21 July 1973, Ley de Minas, Separata del *Boletín Oficial del Estado.*

Gaceta de Madrid (1978), Ley y Reglamento de Minas, Separata del *Boletín Oficial del Estado.*

Observatoire des Stratégies Industrielles (1990), "Les principales caractéristiques de l'industrie espagnole", *Bulletin de la Conjoncture*, January-February.

PEREZ, César L. (1980), "La Legislación Minera Española", *Proceedings of Primeras Jornadas Mineras sobre el Panorama de los Recursos Geológicos-Mineros de Galicia*, Santiago de Compostela, June.

Ministry for Industry and Energy (1981), "Plan Nacional de Abastecimiento de Materias Primas Minerales no energéticas", August.

UNCTAD (1989), *Spanish policy for the promotion of technological innovation in industry: the Centre for Technological Development in Industry (CDTI)*, UN, Geneva.

SWEDEN

INTRODUCTION

Sweden has a long mining tradition and has for centuries been an important producer and exporter of ores and metals in Europe. Although there have been important structural changes in the mining industry, Sweden has kept its position as a major ore producing country in Europe.

The Swedish mining industry is mainly concerned with base metal ores. Mining accounts for about 1 per cent of the market value of Sweden's total industrial production and employs less than 1 per cent of the total industrial labour force; in 1987, the minerals industry accounted for approximately 3 per cent of Sweden's GDP (90 per cent coming from metals and 10 per cent from industrial minerals). Mining production consists principally of iron ore and polymetallic sulphide ores containing copper, lead, zinc, gold and silver.

Non-ferrous ore production has increased in ten years by more than 50 per cent. Most of this increase can be attributed to the Aitik mine at Gällivare which is one of the biggest copper and gold mines in Europe. Alone, it produces about 60 per cent of the Swedish base metal ore supply. The main mining districts are Kiruna-Malmberget (iron, copper), Skelleftea (base metals, gold) and Bergslagen (base metals). Investments in rationalisation and mechanisation made it possible to keep 24 mines in operation in 1989, a number which has remained fairly constant.

Boliden Mineral AB is Sweden's largest producer of copper, lead and zinc concentrates. Its 15 base metal mines are supplying 90 per cent of total Swedish base metal production. Most of the base metal mines are small underground operations. Larger underground mines are Laisvall, Viscaria, Zinkgruvan and Kristineberg. Only copper and lead concentrates are smelted domestically. Sweden's smelter capacity for copper exceeds mine output; imports make up the balance. There is no zinc smelter in Sweden and concentrates are exported to neighbouring Nordic and certain EC Members. Sweden is a net exporter of lead concentrates. The Laisvall and Zinkgruvan lead mine production exceeds the Swedish lead smelter capacity and the balance is exported basically to EC countries.

In the 1980s some gold mines opened. This is the case of Boliden Minerals AB, which expanded gold production in three mines, two of which are copper-zinc mines, and one gold mine with copper and silver as by-products. LKAB (Loussavaara Kiirunavaara AB) — Sweden's major iron ore producer producing and exporting some 90 per cent of Swedish iron ore — conducts exploratory drilling for gold. Terra Mining AB has recently opened up a gold deposit in northern Sweden.

The mining sector is less important for the Swedish economy today than it used to be. The mining industry has played an important part in regional economic development and in many regions, mining together with forestry are still the only significant source of viable economic opportunities. It has also been important for the mining equipment industry.

The Swedish Government is engaged in promoting the mining sector and the objectives of its mining policy can be summarised as follows:

- to provide geoscientific and technological information;
- to promote the search for, and development of, mineral deposits in order to maintain and broaden the domestic supply of mineral commodities;
- to promote the search for minerals, considering geological conditions, in regions with few alternative viable economic opportunities;
- to support, together with the industry, the research and development of exploration, mining and mineral processing techniques and materials technology.

PROSPECTING AND EXPLORATION

Geological services

The Geological Survey of Sweden (SGU) is the central administrative authority for matters concerning the geology of Sweden. Founded in 1858, the SGU is under the jurisdiction of the Ministry of Industry. Its main aim is the mapping of bedrock and quarternary deposits in the country. The main duty of the Geolocigal Survey is to develop, synthesise and market qualified background information concerning the three natural resources: bedrock, soil and ground water. The annual budget of SGU is around Skr 100 million, with a permanent staff of 230 persons. SGU performs geological mapping for mineral exploration and prospecting and for other purposes. The agency also handles questions with regard to legislation on minerals, including those relating to the continental shelf and regulations concerning foreign acquisitions of mining rights.

One on-going programme of interest is the regional mapping of bedrock, started in 1982. The maps are primarily designed as support for mineral exploration, but they can be used also for bedrock geological information. Another programme involves the regional mapping of quarternary information. The original purpose of this kind of map was to satisfy the explorers' need for quarternary information. With regard to these maps, geological information is now computerised.

Lands availability

Three Acts comprise Swedish minerals legislation: the Mining Act, 1974, for ores; the Act concerning Certain Mineral Deposits, for energy minerals; and the Act on the Continental Shelf. Special regulations have been issued for applying all three Acts. Any mineral substance not covered by these three Acts is, in principle, the property of the landowner.

The Mining Act, 1974 is based on the mining claim system. The basic idea in this system is that, in principle, anyone who discovers a mineral deposit may, on application, acquire the sole right to utilise the deposit. The state is entitled to a half-share (state share), in accordance with the provisions in the Mining Act, unless the State relinquishes its right at the staked claim proceedings.

The state share may be transferred or leased, entirely or in part. The State Mining Property Commission is the body responsible for commercial administration of the state mining property and decides on questions relating to the relinquishment of a state share or the transfer or leasing thereof.

A new mineral law passed by Parliament in December 1990 has replaced the Mining Act and the Act concerning Certain Mineral Deposits. An important number of qualified industrial minerals including diamonds are covered by the mineral law, which means that they no longer are considered the property of the landowner.

There are special rules restricting mining operations in most areas of the Swedish part of the Scandinavian mountain chain. Mining operations are only allowed in exceptional cases.

Authority for exploration permits; granting concessions

An exploration permit is issued by the Mine Inspector. It is valid for three years and may be extended for another seven years. If the prospector is successful and finds ore and if he can prove that claimable minerals suitable for technical processing are present in such quantities that mining operations can probably be undertaken, he will be granted a staked claim. Any person may acquire the right to explore mineral deposits on his own land or on land belonging to another.

The claim holder must pay a fee (the claim fee) to the state according to principles decided on by the government and to provide compensation to the landowner or whoever has the usufructuary rights or easements relating to the land. The owner of a property must also be compensated for any damage. In determining amounts, terms of the Expropriation Act are applied.

If the claimant wishes to retain his rights he must apply for exploitation rights (staked claim).

A mineral deposit shall be claimable if it contains:

— gold, silver, platinum, mercury, copper, lead, zinc, manganese, iron, chromium, cobalt, nickel, titanium, vanadium, molybdenum, tungsten, tin, bismuth, antimony or arsenic.

But it is not so, if it is made up of: lake ores or bog iron ores, or contains pyrite, pyrrhotite, graphite, apatite or magnesite.

Tax incentives

In practice, exploration or prospecting expenditures are regarded as part of business costs and are therefore deductible against income from mining.

Direct participation

The government is directly engaged in exploration through the State Mining Property Commission (NSG). NSG's annual budget for exploration is about Skr 50 million. This can be compared with the total average annual expenditure on exploration in Sweden of about Skr 140 million. The State Mining Property Commission carries out exploration through various consultants and contractors. The Commission is also responsible for the commercial administration of the state mining property and mining rights held by the state.

NSG's main exploration targets are base metals, gold and qualified industrial minerals. An important element of NSG's exploration strategy is to broaden the financial and technological base by inviting private industry in Sweden and abroad to form joint ventures. Joint ventures are of special interest for the exploration of minerals which have not been mined to date in Sweden, such as the platinum group elements.

In order to implement its objectives, the government has, in addition to financing NSG and the geological survey, financed an eight year programme (1983-1990) of support of projects for exploration, and research and development, with an annual budget of about Skr 40 million.

The state is also engaged in the mining sector as the owner of the Swedish Geological Co. (SGAB). the SGAB is a consultancy company in the field of geoscience. apart from domestic activities, swedish geological participates in different projects abroad.

MINING AND METALS PRODUCTION

Regulations

When a concession to exploit an area is obtained, the agent must make appropriate payments to the landowner. The state is granted a half-share in all mining concerns. Without exception, since the law was introduced in 1974, the state share has been leased to the owner of the other 50 per cent, against a royalty, normally based on the profit of the mine operation.

A staked claim may wholly, or in part, be acquired by compulsory means in the case either of substantial public interest in its exploration and exploitation or when substantial economic advantages may be gained from its exploitation. Compensation for a staked claim that is acquired by compulsory means may be paid in the form of an annual rent if the size and the nature of the deposit cannot be assessed with sufficient certainty, unless there is an adequate basis for determining a lump-sum payment.

The rent is in the form of a fixed charge plus a production charge calculated on the basis of the quantity or value of minerals produced each year. The rent lasts for 20 years, after which a party may request a revision of the basis for determining the rent.

Before 1938 it was the landowner who was entitled to this half-share but usually the mine owner became the sole owner of all rights by acquiring the landowner's share and compensating him with a royalty equal to 1 per cent of the value of all minerals mined. The royalty had an annual limit and lapsed after 20 years of continuing mining. In addition there was an annual rental fee per hectare which was divided between the landowner and the state. If the mining concern failed to pay these fees, the concession was forfeited.

Minerals taxation

Significant in the case of the Swedish tax system have been the fairly high tax rates and a very narrow tax base. A new tax system was introduced in July 1990 and applied in 1992. The new income tax rate is 30 per cent for all corporations. All income, including nominal capital gains, will be included in one single tax base. Generally untaxed reserves will no longer be allowed. Inventory reserves and similar reserves will be abolished. However, a new tax equalisation reserve will be allowed in the new system. The maximum allocation to this reserve is 30 per cent of equity or 15 per cent of the payroll. Losses can be carried forward for an indefinite number of years.

There are only a few rules applying specifically to the extractive industries. For development expenditure incurred after a commercial deposit has been proven, deductibility of costs for machinery, surface constructions and buildings follows the rules applicable to all business. Costs for machinery are deductible as depreciation with either 30 per cent of the book value or 20 per cent of the acquisition cost. Costs for surface constructions and buildings are generally deductible on a linear basis by 2-10 per cent a year.

All costs for underground work, with the exception of excavations for shafts and main levels, are deductible in the same year as the cost is incurred. This practice has been challenged by the fiscal authorities in some cases. However, the administrative courts have rejected tax authorities in this matter.

The Local Income Tax Act has a rule covering minerals depletion. A deduction may be made as a depreciation over those years the mine is estimated to be in operation.

No special rules apply for close downs or environmental expenditures. Such expenditures can arise from terms of the concession and are normally deductible in the year the cost is incurred. However, if the costs can be estimated in advance, the inclusion of an anticipated cost in the base of depreciation is allowed.

Direct and indirect support

The state is engaged in the mining industry as the owner of the iron ore mines and as owner of LKAB, the large iron ore mining company. The most important mine is Kiruna, operated by LKAB. The government also holds staked claims leased out to private mining companies, namely in the non-ferrous and metals industry. All base metals operations are under the control of private firms.

Other types of intervention concern those related to the rationalisation of the mining sector. One example is the purchase of Grångesberg Mine from Svenskt Stål (SSAB) for 1 krona at the end 1986. The deal was made retroactive to 1 January in order to transfer to the state the losses incurred during the year. The government action made it possible to keep the mine in operation until 1991, mostly to maintain employment.

HORIZONTAL POLICIES

Regional policy incentives

Regional policy in Sweden includes possibilities to support the development of enterprises. Though mining is not included amongst the branches of industry that can be aided regularly, it is possible under certain circumstances to support enterprises in this sector.

Regional aid is restricted to the most northern part of the country where the population density is very low and the distance to larger markets is long. The aid areas cover roughly 59 per cent of the total area of Sweden, but only about 8 per cent of Sweden's population. One of the main conditions for receiving regional aid is that the business is foreseen to be commercially sound.

There are different forms of aid related to investment and employment. The most important one is to support investment in buildings and machinery. The size of the grant is decided in each case after an assessment is undertaken, the main condition being that the aid must not be greater than what is needed to fulfil the planned investment. A maximum share of 35 per cent of the costs of the investment is allowed in areas with the highest priority and 20 per cent in other designated areas.

Even though mining is no longer a widespread business in Sweden, the fact that the major mines are located in designated aid areas makes the branch important from a regional point of view. For this reason, some support has been given during the last years.

Foreign investment controls

Without special permits, foreign nationals or foreign legal entities may not claim mineral deposits in Sweden or work claimed mineral deposits. Matters related to these permits are decided by the government and are regulated in the Act on Certain Limitations on the Right to Acquire Real Property. All applications which have been considered during the 1980s were approved by the government. Currently, the law regulating the right for foreign firms to acquire Swedish companies or mining properties is undergoing revision.

Health and safety

The National Board of Occupational Safety and Health is charged with the regulations for safety and health and imposes limits on lead at places of work. The limit value (occupational exposure limit) for lead is 0.1 mg/m^3 (total dust) and 0.05 mg/m^3 (respirable dust).

The limit for lead in blood is 3.0 µmol/l (= 62 µg/100 ml), which means that a worker must be removed from lead work if his blood sample contains a higher level than indicated. Also, if the blood sample exceeds 2.5 µmol/l of lead in three consecutive samples (interval 3 months), he must also be removed from lead work, until the blood sample shows less than 2.0 µmol/l.

Environment

The Environmental Protection Act (1969) is concerned with pollution and other environmental disturbances caused by permanent plant. Certain types of industries and other installations may not be established or modified without a licence, as is also the case for the discharge of certain effluents.

Before a mining operation can be started, an environmental concession must be granted. The concession regulates that the operation is in compliance with the Environment Protection Act.

According to the new mineral law which was scheduled to enter into force on 1 July 1992, replacing the Mining Act and the Act concerning Certain Mineral Deposits, the Act concerning the Management of Natural Resources will be applied in matters dealing with concessions for mining operations. One section of this Act stipulates that land, water and the physical environment in general shall be used in a manner that encourages good long-term management at ecological, social and economical levels.

Licence applications are examined by the National Franchise Board for Environment Protection and the respective county administration. When granting licences, they stipulate criteria such as maximum levels of emissions, measures required to protect the environment and so forth.

Before a concession is granted, conflicting interests are considered. This means that in matters dealing with concessions for mining operations there must be an arbitrage between the interest to start the mining operation and those that are opposed. Examples of potentially conflicting interests include environmental protection, agriculture, forestry and reindeer husbandry. It must be said that the interest to extract mineral products is an interest mentioned in the Act and thereby enjoys the same rights as other interests mentioned in the Act.

The authorities are also responsible for supervision, which involves tracking down cases requiring action, monitoring sources of environmental disturbance, enforcing protective regulations and prosecuting breaches of the Act.

The following is one example of an environmental project which has affected non-ferrous activities, through its objectives of eliminating the causes of water pollution in the Dalälven river, preventing similar pollution in the future and reducing the off-load of environmentally harmful substances into the Baltic. The Dalälven Commission was appointed by the government and recently presented its report. The Commission set its targets for the greatest possible reduction of all discharges of pollutants — chiefly heavy metals, oils, other chemicals and nutrient salts — into the Dalälven river. Since heavy metals are discharged mainly from deposits of mine wastes, the Commission decided to assign the highest priority to measures aimed at limiting such discharges. It was found that numerous stockpiles of mine wastes are present in the Dalälven river's run-off area and that weathering leads to subsequent leaching of metals. Several of the deposits are in themselves ancient remains and relics, or are related to such, and in many cases they are of major historical value, in certain cases of national interest. The City of Falun has been proposed as a Protected Area on the UN World Heritage List. Assigning priority to deposits for clean-up and selecting the most suitable measures, necessitated the collection of extensive background data.

At the end of 1990, Boliden was appealing the stringent proposed environmental regulations, particularly because meeting them would significantly curtail copper and lead output from the Rönnskär smelter. The company developed a US$ 10 million programme to keep its smelter open while it reduced sulphur dioxide (SO_2) emissions by 50 per cent. The plan was to close several outdated copper and lead production lines as well as sulphuric acid and liquefied SO_2 plants. Except for sulphur dioxide, production levels would be maintained through efficiency improvements and capacity increases in newer lines and plants.

Research and development

Co-operative technical research related to the mineral and metal industry is jointly financed by the government via The Swedish National Board for Technical Development (STU) and by the companies concerned. This board is the central agency for basic research and applied research in the field of technology as well as industrial technical development. The co-operative research related to the mineral sector is carried out at research institutes and institutions of higher education. Areas of research are ore geology related to prospecting, mineral processing techniques, mining technology and mining equipment, detonation technology, industrial minerals and materials technology.

Research projects of specific interest for the non-ferrous ore industry are:

— "Ore geology related to prospecting" (1988-1994) financed by STU for Skr 10 million, Ministry of Industry Skr 8 million and mining companies Skr 6 million. During the first three years of the six year programme research has been carried out mainly on volcanogenic massive sulphide deposits in the mining districts.

— "Chemistry of sulphide Mineral Processing" (1985-1991) financed by STU Skr 13 million and the Ministry of Industry for Skr 5 million. The aim is to develop cost efficient process technology for dressing of fine grained sulphide ores.

— "Mining Technology 2000" (1987-1992) financed by STU for Skr 50 million and the mining and mining equipment industry Skr 200-300 million. The aim is to develop more efficient underground mining methods.

The following research institutes are carrying out research related to the minerals and metals industry.

- Swedish Detonic Research foundation
- Swedish Institute for Metals Research
- MEFOS — the Foundation for Metallurgical Research
- Swedish Mineral Processing Research Foundation
- Swedish Rock Engineering Research Foundation

Inputs

Energy policy is also the subject of important discussions in Sweden. In January 1991, an agreement on energy policy was reached by the Social Democrats, the Liberal Party and the Centre Party. A government bill based on the three-party agreement was accepted by Parliament in June 1991. The stated purpose of the agreement is to secure the long-term and short-term supply of electricity and other energy on internationally competitive terms, thereby furthering good economic and social development in Sweden. A strategy for reducing climatic impact is included in the decision. Energy policy must be based on what is naturally and environmentally sustainable. National work to combat climatic change is organised in such a way that Sweden, jointly with other Western European countries, can give added impetus to work at the international level.

Electricity prices are among the lowest in Western Europe. It is stressed in the Bill that it is the task of the power companies, guided by energy policy laid down by the government and the Parliament, to plan and develop the power system so as to meet national electricity demand. Given the expectations for a satisfactory electricity balance, increases in electricity prices during the first half of the 1990s are expected to be moderate.

Trade policy

Sweden maintains very low tariffs on the imports of all six non-ferrous metals. With the exception of a few chemicals (dutiable at rates of 3.8 or 5.8 per cent, although often benefiting from applied rates of zero), the vast majority of metal imports, particularly for ores and unwrought metal enter free, with the remaining tariffs on wrought metal and finished manufactures ranging between 1.4 and 3.8 per cent. Under preferential arrangements, developing countries benefit from free entry for MFN dutiable products in the wrought metal and finished manufactures categories, as do other European Free Trade Association (EFTA) countries and EEC Members. Hence the MFN rates have in fact applied to imports from Japan, Canada, the US, and eastern European countries. Based on trade flows in 1983 and 1984, the GATT calculated actual dutied imports as representing between 0.2 and 8 per cent of total Swedish imports for the six non-ferrous metals.

CONSUMPTION/SUPPLY

Stockpiles

Swedish preparedness, including adequate supplies of raw materials, depends on maintaining a multi-faceted and flexible industrial fabric which, with the help of limited, supplementary measures, can satisfy both civil and military needs in different crises. According to the 1987 decision on defence which defined the national understanding of a so-called peace time crisis, national stockpiles of certain metals and destined for use in a peace time crisis should be gradually set up. In the future, it is the task of companies, through their commercial stockpiles, to be responsible for civil preparedness, i.e. by being capable of meeting peace time disturbances in supplies.

Whilst the state National Board of Civil Emergency Preparedness (ÖCB) does not administer stockpiles for peace-time crisis, it has contributed to developing a very special measure of preparedness — the so-called standards of preparedness for the most important types of commercial steel and stainless steel. The purpose is to be able to economise certain alloying elements swiftly. In 1977, Sweden began building up a strategic stock, designed to cope with possible supply disruptions due to political tensions in southern Africa. Swedish production of speciality steels led the country in 1979 to buy chrome and vanadium from South Africa, chrome from Zimbabwe and cobalt from Zaire and Zambia equivalent to one month's consumption.

Recycling

A recycling system for aluminium cans was introduced in Sweden in 1984. The recycling rate has been raised from 60 per cent to 83 per cent over the last six years. In accordance with a recently passed law, the return rate for aluminium beverage cans should reach 90 per cent by 1993. A special company, AB Svenska Returpack, is operating the recycling system. The system is based on mandatory deposits with profits allocated for environmental purposes.

In December 1990 the Parliament passed new legislation on a system for the recycling of lead batteries in Sweden. Since 1 January 1991, lead batteries carry an environmental cost charge of Skr 32. The charge is transferred to a special battery fund, which is administered by a government agency. The funds are to be used for information campaigns and as contributions to collection costs.

The aim is to reach a collection rate of 95 per cent within two years. The three branches concerned in Sweden — the scrap trade, the battery producers and the smelter industry — have formed a joint company, Returbatt AB. Returbatt AB provides information about the system and co-ordinates collection.

Boliden has striven to integrate more fully its recycling activities within the Nordic markets before searching to reinforce its position within the EC. In the late 1970s Boliden bought two subsidiaries of the Danish group Bergsöe to control the Scandinavian lead market. More recently, Boliden bought 90 per cent of the Swedish Paul Bergsöe & Son that controls the Finnish Oy Bera. These two firms recuperate lead from recyclable material equivalent to one half of Boliden's production. As a consequence of this operation Boliden will partially control the Swedish and Finnish markets. Boliden's reinforcement in this segment lead the Swedish firm to take 10 per cent at Bergsöe that controls several smelters around the world jointly with the group Det Ostasiatiske Kompagni.

UNITED KINGDOM

INTRODUCTION

Production in the United Kingdom of all base metals is far below its consumption. The major part of UK consuming industry requirements are met by imports, the remainder coming from domestic production and recycling. Based on 1986 statistics, the report "Mining in the UK — The Industry and its Competitiveness"[1] stresses that there is a 100 per cent net import dependence for nickel, titanium minerals, gold and silver; 80-90 per cent for aluminium, copper and zinc. Dependencies for lead are shown as 39 per cent, and tin as 25 per cent.

About 7 per cent — some US$ 35 billion — of 1987 GDP was accounted for by the minerals industry, including fuels, metals and mineral products. The processing of non-ferrous metals amounted to about 20 per cent of this total but the contribution due to mine production of ore or concentrates was small, perhaps US$ 30 million. Of this, the mining of tin concentrates, iron ore and zinc concentrates accounted for most. The exploitation of tin has been the most important mining activity in the UK. After the end of International Tin Agreement buffer stocking operations in 1985, and the consequent fall in prices, the British tin industry experienced a very difficult period with a decrease in mine output.

Today the only producer is Carnon Consolidated Tin Mines through the South Crofty and Wheal Jane mines. The latter was due to close at end 1990, but won a reprieve to remain in production until mid-1991 because of good prices for its by-product, zinc. But in March 1991, Carnon announced the immediate closing of the two mines, due to a decision by the government to stop payments on its support grant, though the South Crofty mine has since reopened at a reduced output. Low tin prices had already forced the closure of the other producer, Geevor plc in 1990.

In evaluating the mining industry's contribution to the UK economy, one must also take into account the country's position as a major home for international mining companies, and the fact that the mining houses bring in major inflows of interest, profits and dividends. Historically, the importance of British mining companies and of industrial consumption account for the location of the world's major metal market, the London Metal Exchange, in London's City.

More recently, the participation of the British mining houses in overseas mining projects with Japanese, North American or European partners in order to spread risks and pool technologies and financing gives them a particularly strong position in world markets. Thus, for instance, after the acquisition of BP Minerals in 1989, Rio Tinto Zinc (RTZ) became the biggest mining company in the world. The deal raised RTZ's share of world copper production to about 5 per cent and this will increase to nearer 7.5 per cent when new projects now in development or advanced feasibility stages come on stream. The company has also significant shares in world production of silver, lead, zinc, tin and gold. It now operates in 40 countries with 90 per cent of its earnings derived from outside the UK. The principal areas of interest are Australasia and the USA, but RTZ is developing its mining activities as far afield as Brazil, Chile, Spain, Portugal or Zimbabwe.

Despite past industrial policies and the government's intervention in some basic industries, there is no specific policy for non-ferrous metal mining or processing in the UK. These industries are subject to general legislation in various fields such as planning, health and safety, taxation, employment and trade and industry. In contrast, British regional policy has remained fairly active over the years. With few exceptions, the areas with the worst economic problems are those in which mining and mineral processing activities once played an important role in the regional economy and today benefit from promotional measures taken under regional assistance programmes.

The basic laws regulating mining activities in the UK are the Mines and Quarries Act of 1954 and the Mines (Working Facilities and Support) Act of 1966, as amended in 1974. Environmental legislation was established by the Town and Country Planning (Minerals) Act of 1971 and 1972, as amended in 1981, and taxation rules applicable to mining are contained in the Capital Allowance Act of 1990. The UK non-ferrous metals industry is subject to a whole host of environmental and activity related legislation, not least of all the Control of Pollution Act 1974 and, more recently, the Environmental Protection Act 1990. The Department of Trade and Industry (DTI) (formed by the merger of the previously separate Departments of Trade and Industry) is the interface between government and the metalliferous and industrial minerals mining industry.

PROSPECTING AND EXPLORATION

Geological services

The British Geological Survey (BGS) — previously called the Institute of Geological Sciences (IGS) — is the main governmental agency for undertaking work in earth sciences and is the recognised depository for geoscience data. In Northern Ireland, the geological survey is carried out by the Geological Survey of Northern Ireland, funded by the Department of Economic Development (DED).

Since its foundation in 1835 the BGS's primary function has been the geological survey programme, which includes the UK continental shelf. The BGS also operates in overseas countries on behalf of the Overseas Development Administration, and participates in international research on the deep sea bed.

The UK land survey and the continental shelf study have both been carried out by the BGS, supported partly by the Department of the Environment and (offshore) by the Department of Energy. Minerals surveys, carried out through programmes financed by the Department of Trade and Industry, are as follows:

- The Mineral Reconnaissance Programme (MRP) began in 1972 and is a nation-wide programme of investigations designed to identify new areas with potential for metalliferous and related minerals in order to obtain information likely to be of assistance to private sector exploration companies.
- The Geochemical Survey programme (GSP), formerly the Regional Geochemical Reconnaissance Programme (RGRP), undertakes the systematic collection of geochemical data from the whole country based on high-density sampling of stream sediments, with analyses for up to 30 elements. The MRP and GSP, both of which are totally funded by the Department of Trade and Industry, are establishing a database of national importance both for mineral exploration and environmental studies.
- The Mineral Intelligence Programme (MIP), largely funded by the Department of Trade and Industry, provides information and authoritative advice on economic minerals and many aspects of the minerals industry at the international level (world mineral resources, production and trade) and maintains a database on the mining industry in Britain.
- Mining companies' exploration data. Companies and landowners may hold the results of past exploration but there is no central registry for such information except for exploration activities assisted by the Mineral Exploration and Investment Grants Act 1972. Under this scheme grants were given on condition that the results were deposited with the BGS. In Northern Ireland all exploration data are deposited annually with the Geological Survey of Northern Ireland, but upon request they can be kept confidential for a period of five years or more.

The BGS also has an active programme of geophysical investigation, as well as undertaking mineral resource evaluations and environmental studies.

Lands availability

In Great Britain, the broad framework for the use of land is provided by Structure Plans that cover most parts of the country. Mining prospectors must respect these development plans that often contain useful information on past and current mineral working in the area, policies for future mineral development and criteria for the restoration and after-use of mineral sites.

All mineral working activities come under the control of the Town and Country Planning Acts, but activities with no impact on the environment, such as preliminary prospecting, do not require specific planning permission. Various local planning authorities have powers to grant or deny building permits and through these attributions they arbitrate among competing interests for land use.

Applications for mineral working in some environmentally sensitive areas such as National Parks and Sites of Special Scientific Interest will be subject to the most rigorous examination. However, large scale mineral operations for china clay, fluorspar, limestone, potash, roadstone and slate are carried our within National Parks from open pit and underground mines on a basis of national need and local employment. In practice, it is established policy that the Secretary of State for the Environment will only exercise his right to "call in" minerals planning applications if they raise issues of regional or national importance.

Even though such planning permission procedures cannot be regarded as deliberate restrictions on exploration of mining activities, in practice they may effectively limit the access of mining firms to land. In 1977, a government report (the "Stevens Report") urged changes in planning control over mining operations. In 1978, the Department for Industry stated a general interest in promoting the development of mineral resources, but recognised that competing land-use requirements had to be reconciled on a case-by-case basis. There is a presumption against development in the following areas: the highest grades of agricultural land (except in the case of silica sand), the built heritage, important archaeological remains and their settings, Ministry of Defence land, sea defence land and inappropriate development in Green Belts.

Authority for exploration permits and granting permissions

In Great Britain the mineral rights to non-fuel minerals, with the exceptions of gold and silver, are mainly in private ownership, although a significant proportion is owned by the Crown and by government departments and agencies. The surface rights are often separated from the mineral rights. In contrast, in Northern Ireland, most minerals are vested in the state and the Department of Economic Development is responsible for granting exploration and mining licences.

Gold and silver are a monopoly of the Crown. Their exploration and development must be done under a licence given by the Crown Estate Commissioners and the Crown Mineral Agent. However, Crown Estate permits, licences and leases confer no right of entry and the applicant must negotiate surface access separately with the surface rights owners. Since 1987 the results of prospecting for gold and silver have been deposited with the BGS by the Crown Estate Commissioners. The British system however presents a very important complexity: in fact, even if the general rule is that the mineral rights are held by the surface landowner, in areas of a long history of mining such as South-West England there are many cases where the mineral rights have been retained by a previous landowner when the surface freehold was sold. In fact there is not a national register of mineral rights in Great Britain, even if some details about surface ownership and current ownership of mineral rights can be found in the Land Register.

The search for the title-holder of the mineral rights can be a long, arduous and expensive task[2]. Once the owner of the mineral rights is identified, the usual way to obtain research and exploration permits is through direct negotiations between the mining company and the owner of the mineral rights.

The Mines (Working Facilities and Support) Act 1966, as amended, provides a means by which an operator can obtain the necessary authority to explore for and work minerals, in case of difficulties in identifying the mineral rights owner.

In Great Britain, to move from exploration to mining development requires planning permission. This application by the future developers is considered by a planning committee formed by elected councillors advised by the County Planning Officer and assisted, in mineral cases, by the Minerals Officer. The planning framework in England and Wales is provided by the Town and Country Planning Act 1971 as amended by the Town and Country Planning (Minerals) Act 1981. Day-to-day responsibility for the administration of the planning system is in the hands of mineral planning authorities, mainly the county councils.

Obtaining planning permission is the most difficult stage. Several statutory authorities must be consulted and local authorities may ask for additional information concerning matters such as the environmental effects of the proposed activities. The requirement of environmental impact studies, previously done on an informal basis, has been regulated in the Town and Country Planning Act 1990, implemented in August 1990. If the application is refused, the applicant has the right of appeal to the Secretary of State for the Environment and this usually results in a public inquiry, conducted by a planning inspector. This process can take as much as five years as it is conducted through a series of informal and formal discussions involving the local communities with the aim of achieving a solution reconciling the interests of all the parties involved.

In Scotland, the relevant Act is the Town and Country Planning Act (Scotland) 1972, as amended by the Town and Country Planning (Minerals) Act 1981. There is no separate regime for mineral planning, proposals for mining activities being dealt with by the authority responsible for all forms of development. The responsibility lies with the District Council and Regional Councils.

In Northern Ireland, planning permission is not usually required for normal prospecting activities. However the Department of the Environment in Northern Ireland is responsible for planning and is informed when a prospecting licence is issued.

A recent report on the UK Mining Industry and its Competitiveness[3] calls for government intervention to reduce the complexity and cost of obtaining access to mineral deposits, which could increase the competitiveness of existing producers and help reduce obstacles to the entry of new producers.

Grants and tax incentives

Assistance to UK firms (domestic or foreign-owned) for private exploration within the United Kingdom existed from 1971 to 1984 under the Mineral Exploration and Investment Grants Act 1972 (MEIGA), under which the Department of Trade and Industry refunded up to 35 per cent of exploration costs for approved programmes. Following a review of the scheme, the Secretary of

State for Trade and Industry suspended any further offers of assistance in July, 1984 and the scheme was officially terminated in March 1985.

Grants were restricted to the exploration and evaluation stages of a project. Where the exploration led to the extraction of mineral deposits in commercial quantities within 12 years of receipt of the subsidy, such aid was repayable, with interest, to the government. The Department of Industry had the authority to decide whether a project had merit for purposes of the programme and often required a guarantee or security for repayment[4]. The distribution of these grants corresponded quite closely with the "assisted areas" under the various UK regional programmes. Most of the approved grants were for exploration of non-ferrous metals in Scotland and Southwest England.

Companies carrying on mining activities are liable to corporation tax under the normal rules. If there is trade in mineral extraction, the exploration expenditure is considered capital expenditure and will qualify for relief at 25 per cent on a reducing balance basis. An activity which consists simply of exploring for minerals is not regarded as trade. Trading only commences when the decision to develop a source of minerals is taken. In these cases, the exploration expenditure may be subject to annual allowances in certain circumstances. As exploration is not, on its own, a trading activity, losses arising from this activity cannot be set off against other income. The absence of exploration incentives schemes has been criticised by the Mining Association of the UK as putting the UK at a disadvantage vis-à-vis countries who do so and as favouring the larger companies or those already involved in mineral exploration.

MINING AND METALS PRODUCTION

Mining

Minerals taxation

The government's overall economic strategy to create a climate for business activities has involved reductions in tax burdens on business, among other measures. The standard rate of corporation tax fell from 52 per cent in 1982 to 35 per cent from 1986 and has now fallen further to 34 per cent for the financial year starting April 1990 and 33 per cent for subsequent years. The small companies rate has been 25 per cent since 1 April 1988 on profits up to a certain level, £200 000 for 1990/91. The rate increases gradually on higher profits until it reaches 34 per cent on profits of £1 million (for 1990/91). If a company is associated with other companies the limits of £200 000 and £1 million are reduced. All corporate income, including dividends from foreign subsidiaries and profits from foreign branches, is subject to the same rate. The reduced rates do not apply to branches of foreign companies. There are no special rates applicable to mining companies.

Tax incentives

One of the key measures to promote industrial investment has been the changes in The Mines and Oil Wells Allowances Act implemented in the 1986 Finance Bill. The reduction of the corporation tax has been accompanied by the abolition of accelerated capital allowances for plant and machinery and industrial buildings. For mining, the main effect of these changes was to simplify the system and bring previous capital allowances in line with those of other industries. The previous system of allowances has been abolished and replaced by a straightforward percentage based annual writing down of allowances on a reducing balance basis.

Costs of restoring sites are usually allowed as revenue deduction in the year the expenditure is incurred. The cost of acquiring the minerals will also be relieved by Mineral Extraction Allowances. The annual allowance is restricted to the cost of the minerals — excluding the value of the undeveloped land. The cost of the minerals is allowed at 10 per cent a year on a reducing balance basis.

Redundancy payments made by virtue of a contribution received from an approved industrial rationalisation scheme will normally qualify as a deduction from profits under Section 568 ICTA 1988, to the extent that they are not allowable under any other provision.

Allowances on machinery and plant are available at the rate of 25 per cent a year on a reducing balance basis from the date when the machinery was bought. Industrial building attracts annual writing down allowances on a straight line of 4 per cent of its original cost.

Capital expenditure on scientific research (excluding the cost of land and houses) related to a company's trade can be completely written off against tax in the year of the expenditure.

Subsidies

In the autumn of 1985, the Cornish tin industry had requested £50 million to be injected over five years to modernize the tin mines, and enable them to compete in the world market. In June 1988 RTZ Metals sold off Carnon Consolidated Ltd. to Carnon Holdings Ltd., a company owned by the management (80 per cent) and the employees (20 per cent). At that time, RTZ provided Carnon with a interest-free loan of £10 million and the Department of Trade and Industry contributed with a loan of £25 million, interest-free. One of the conditions placed on Carnon was that any income from other sources must be used to support and sustain tin mining. With continuing low tin prices, DTI decided in February 1991 that it would no longer continue disbursing the £25 million loan; RTZ also withheld their loan. Carnon immediately closed Wheal Jane (which was due to close in 1992 anyway) and commenced care and maintenance preparation for South Crofty. Subsequently, however, they have continued to run South Crofty with a scaled down workforce on short-term contracts.

Metals production

Inputs

No general programme exists relating energy policy to the non-fuel mineral sector, especially to energy-intensive industries like aluminium smelting. In principle, prices of electricity are set according to long run marginal costs of supply, guided by world market prices.

Anglesey Aluminium and British Aluminium Co. (now British Alcan) both abandoned their plans for capacity expansions in the late 1970s because cheap energy was no longer available. In 1981, British Aluminium decided to close its Invergordon smelter built in the late 1960s. It became an unprofitable operation with rising power costs — the smelter's long term power cost had been based upon an advanced gas-cooled reactor nuclear station — in a situation of falling aluminium prices.

Invergordon smelter was supplied from Hunterston B nuclear plant. Government loans were extended to Invergordon to finance its share of the nuclear plant against a long term supply contract until the year 2000. It further benefitted from subsidies, the government having over the years paid out over £100 million to compensate the electricity authorities, because Hunterston was completed three years late and delivered much less electricity than had been envisaged under the terms of British Aluminium agreement. For three years, Invergordon was supplied by the North Scotland Hydro-Electric Board (NSHEB) with much more expensive electricity drawn from the national grid. In fact, problems with the design of the station forced it to operate at less than half the expected capacity and the increase in the per unit electricity prices could not be passed on to British Aluminium under the terms of the contract.

The Anglesey Aluminium smelter [Kaiser (49 per cent) and RTZ (51 per cent)] had the same sort of problem because the nuclear power station at Dungeness started operating ten years late. As a consequence, Anglesey was also supplied by other units on the national grid. In the early 1980s, it cut three-fourths of the smelter's capacity.

Direct and indirect support

Mineral processing industries may be eligible to benefit from various forms of aid from the central government or other bodies.

With the closure of the Regional Development Grant scheme in 1988, public assistance is virtually all in the form of Regional Selective Assistance (RSA). RSA is available under the Industry Act of 1982 for investments throughout the assisted areas. The purpose of the scheme is to reduce the imbalance in employment opportunities in the Assisted Areas by encouraging investment projects which would otherwise not proceed in the form proposed.

Grants are negotiated at the minimum level for the project to proceed — subject of course to EC limits. Assistance is available for either job creation in association with new investment or job safeguarding in association with a modernisation.

Applicants located in "assisted areas" may, if they satisfy all the criteria of the scheme (which include the need for assistance, viability of the applicant and — with the assistance — the project, and that the project brings both regional and national benefit) receive grants toward all types of investments. Applications are vetted to ensure they comply with EC sectoral restrictions.

Since the mid 1980s, the UK Government has provided neither Exchange Risk Cover nor guarantees on new loans for EIB. The UK Government, however, still has some remaining commitments on EIB loans made before the mid 1980s and continues to provide exchange risk cover on ECSC conversion loans.

Government ownership

Several industrial sectors in the United Kingdom have, in different periods, been subject to government ownership. Thus, for instance, the coal and steel industries were nationalised and the National Economic Board (NEB) made several equity investments in the automotive and electronic industries.

The only case of state ownership in the mineral processing industry is the financing of a titanium convertor plant designed for production to supply Rolls Royce. The Industrial Reorganisation Corporation, predecessor to the NEB, provided a US$ 1.2 million loan to the Davy Ashmore Aluminium refinery to cover start-up costs.

HORIZONTAL POLICIES

Environment

The House of Commons and the House of Lords recently approved the Environmental Protection Bill 1990 which received Royal Assent on 1 November 1990. The Act seeks to establish an integrated system for pollution control. This is the latest in a series of environmental statutes originating from the nineteenth century (the Public Health Act 1848, the Public Health Act 1875, Public Health Act 1936, amplified by the Health and Safety at Work Act 1974, Control of Pollution Act 1974 and similar legislation), the mining regulations having been updated by the Health and Safety Commission in order to replace the 1954 Mines & Quarries Act and the 1971 Mines Management Act. The first set of new regulations came into force in April 1988.

On the other hand, the Planning Legislation, which now regulates all forms of economic development including mining, has recently been consolidated in the Town and Country Planning Act 1990 which was implemented in August 1990. This Act requires that the consent of the Local Planning Authority be obtained for the carrying out of all forms of development. The Planning Authority is empowered to impose planning conditions, which in the case of opencast mineral operations, include monitoring and controlling dust emissions, noise and vibrations.

One group of Parliamentary statutes has been directed towards improving air quality. The regulations began with the Alkali Works Regulation Act 1906, later developed by the Clean Air Acts 1956 and 1968 and the Health and Safety at Work Act 1974.

National subordinate legislation details the regulations of primary Act of Parliament. Since January 1988, policy guidance on topics of general relevance to the work of planning authorities have been issued in the form of Planning Policy Guidance Notes, when dealing with matters of national or regional significance; Minerals Planning Guidance Notes, when dealing with the control of minerals development; and (since mid-1989) Regional Planning Guidance Notes, when dealing with the Secretary of State for the Environment's strategic planning guidance for particular regions.

The Minerals Planning Guidance Notes are: MPG1 (General Considerations and the Development Plan System), MPG2 (Applications, Permissions and Conditions), MPG4 (The Review of Mineral Workings Sites), MPG5 (Minerals Planning and General Development Order), MPG6 (Guidelines for Aggregate Provision in England and Wales), MPG7 (The Reclamation of Mineral Workings Guidance), MPG8 (Mineral Planning Guidance: Index May 1990). The topics they cover can be national, regional or strategic in scope.

The European Community Directives are influencing UK mining regulations. As a consequence, in November 1990 new environmental legislation, the UK Environmental Protection Act, 1990 was adopted. This environment legislation deals basically with "Implemented Pollution Control" (IPC), setting stricter emission standard levels for all industrial sectors and laying down a timetable for implementing them. By the beginning of April 1996, all sectors must be covered.

The metals industry has received the guidance notes in January 1991 but this sector apparently comes low on the list of priorities for ensuing legislation (1994-95). Under the bill, all metallurgical furnaces with a capacity of over five tonnes will be inspected by Her Majesty's Inspectorate of Pollution (HMIP) while those under five tonnes will be controlled by the relevant local authority. IPC autorisation will be required for all non-ferrous plants by February 1995. However, the UK aluminium diecasting and secondary ingot-making capacity will be under financial pressure earlier on because the new legislation brings them under local authority supervision for the first time. Until now these facilities were categorised in such a way as to be exempted from all smoke legislation. It has been suggested that these new mesures may drive small producers out of business (about a third of the UK secondary ingot production capacity) or, at least, remove them from their low cost position.

Research and development (R&D)

The non-fuel mineral sector has not figured prominently in the government R&D effort. However, support programmes can be found directed at the non-fuel minerals sector in several government agencies:

- Department of Environment supports geological mapping activities and mineral reserves evaluations of the British Geological Survey (BGS) and various marine science research institutes which are active in geological and mineral extraction technologies.
- National Research Development Corporation, an independent public corporation supported by the Department of Trade and Industry which provides manufacturing firms with finance for development, prototype production and exploitation of inventions.
- Department of Trade and Industry funds several research establishments which conduct general industrial research directly or through contract.

Finally, companies undertaking R&D are not compensated by any special tax measures although they may qualify for Scientific Research Allowance.

Trade policy and marketing

The British Overseas Trade Board promotes exports of manufactured goods and the Export Credit Guarantee Department guarantees export credits. These measures may have affected the mining sector, albeit marginally.

As a Member of the EEC, the UK applies the common external customs tariff, the section of which relating to the six non-ferrous metals is summarised in the chapter dealing with the European Economic Community.

CONSUMPTION/SUPPLY

Securing overseas supply

Historically, British policy for securing raw materials was through its colonial administration. This permitted support of firms participating in mining activities world-wide and to develop trade and foreign policy ties with the Commonwealth countries. Since colonial times, mining houses such as RTZ or Consolidated Gold Fields have played a major role in creating a flow of raw materials to Europe and the UK, in particular. Even if the political independence of most English colonies contributed to a decrease in control by these companies over production, today they are called on to provide expertise, managerial capabilities and capital for the development of raw material production, and as organisers of raw materials markets.

Support to exploration

BGS overseas programmes are carried out on behalf of, and are mainly funded by, the British Government's Overseas Development Administration (ODA). They are of two main types, one country specific, the other regional in their application and emphasis. The country specific programmes are aimed at training and strenghtening of counterpart organisations. These are technical co-operation (TC) programmes under long-term assignments carried out by residential BGS teams and backed by shorter visits from headquarters-based staff. BGS also continues to second staff to strengthen geological survey departments in developing nations of Commonwealth under the terms of the Overseas Service Aid Scheme (OSAS). The regional programmes are research-oriented and are carried out by UK based staff who go overseas on short-term assignments. They aim to generate effective techniques of mineral, energy and groundwater resource exploration and development that are applicable and will find general use in developing countries.

The BGS is undertaking either TC or OSAS programmes, and in some cases both, in twenty-eight overseas countries. Activities with a strong metals exploration element are currently undertaken in Latin America (Bolivia, Colombia, Ecuador), the Far East (Indonesia, Laos, Malaysia) and Africa (Zimbabwe). Regional research and development programmes total twenty-two in number. They include ammonium geochemistry in the search for hydrothermal gold mineralisation (Mexico and Indonesia), and the development of a standard field kit for geochemical reconnaissance and publication of technical handbooks (not country specific).

Foreign assistance to producers

The UK government does not provide any assistance to national companies producing overseas, even if they ensure mineral supplies for the UK industry. The government considers that British mining houses are in good position to take advantage of mineral investments overseas by the international development banks and well placed to participate in projects in former colonies where their presence is still important.

Tax measures for overseas investments

UK residents may credit their UK taxes or deduct from their income taxes paid to foreign governments on income earned there. The credit is generally limited to the lower of the foreign tax or the UK tax on the same income. The UK marginal tax rate is used for the calculation, irrespective of the foreign tax rate calculation. UK depletion allowances for the acquisition of mineral rights apply equally to both foreign and domestic operations. For overseas activities, in addition to allowances for mining works, further relief is given for facilities to provide services for employees. UK firms are able to write off all abortive exploration spending, foreign and domestic.

Insurance to domestic companies investing overseas

Overseas investment insurance is a scheme administered by the Exports Credit Guarantee Department (ECGD) and is available to UK firms and their overseas subsidiaries. Cover is provided against the main political risks. These are:

- Expropriation — which includes direct nationalisation as well as indirect forms of expropriation by foreign governments which aim to discriminate against UK investors.
- War damage — due to revolution or insurrection.
- Restrictions on remittances, affecting repatriation and currency conversion.

The standard commitment under this scheme runs for an initial 15 years and covers 90 per cent of any loss arising from these types of risk.

The scheme applies to both equity and loan investments. An overall maximum amount of coverage is agreed to by the ECGD and the investor. Premium is charged on the current insured amount at an ad valorem rate ranging between 0.7 per cent and 1 per cent depending on the host country concerned.

At 31 December 1990, there were 102 agreements in force covering 72 projects. These investments, valued at £96.5 million are located in 40, mostly developing, countries.

Stockpiles

The stockpile of strategic minerals was established in 1983 as a contingency against short-term disruption in supplies of a number of minerals critical to certain sectors of UK industry. The Secretary of State for Trade and Industry announced in October 1990 that the stockpile would be sold over the next two to three years. A condition of the sale was that disposals take place in a manner which seeks to avoid undue disruption to the market. Her Majesty's Government owns and finances the stockpile. The Department of Trade and Industry is responsible for its day-to-day management.

Recycling

The UK's National Anti-Waste programme, which began in 1974 is co-financed by the Departments of Trade and Industry and of Environment. It is concerned with the involvement of localities, industry and environmentalists in recycling and it must provide information services, support research and development and preparation of packaging codes or support pilot plants for the mechanical separation of refuse. The goal of the Environment Secretary is that half of domestic recyclable waste should be recycled by the turn of the century. To promote recycling, the Department of Trade and Industry has a research establishment, the Warren Springs Recovery Division and operates a Waste Materials Exchange.

Activities in the field of research and development on recycling are largely expected to be initiated and therefore funded by the private sector. However, central government can and does provide financial incentives for recycling R&D in particular circumstances although the scheme of assistance is not solely targetted at recycling efforts.

The Department of Trade and Industry, in conjunction with the Department of the Environment, operates the Environmental Technology Innovation Scheme (ETIS) which offers support for innovative research in environmental and clean technologies which will lead to improved competitiveness for UK industry and to technologies which pave the way for improved environmental standards.

The Department of Trade and Industry's Environmental Management Options Scheme aims to support the widespread adoption of technologies and best practice techniques with potential for environmental benefit in such as cleaner technologies, waste minimisation and recycling. Because R&D projects with a recycling element are not appropriate to a separate scheme and therefore not separately identifiable, it is not possible to put a figure on UK government's financial contribution to such expenditure.

The UK's current industrial recycling performance is on a par with most other European countries. In particular, the UK can claim notable successes when looking at the percentage of non ferrous metals recycled compared to consumption. In 1989, for example, some 65 per cent of lead and 40 per cent of copper consumed in the UK was recycled metals. The UK performance on domestic waste recycling however — including, for example, aluminium cans — is not so well developed. Estimates for recovery range from 2-5 per cent of all such waste. Therefore in its 1990 White Paper on the Environment the government set a target that by the year 2000, 25 per cent of all household waste would be recycled. In the particular case of aluminium cans, however, the current rate of 10 per cent is likely to improve dramatically once British Alcan's planned remelt facility is commissioned which will have the capacity to process at least 50 000 tonnes of used beverage containers.

The Environmental Protection Act 1990 will have an important impact on the recycling industry. Inter alia it will introduce a system of credits to encourage recycling and reduce the amount of material going for disposal by ensuring that savings in disposal costs are credited to the organisation which removes the material from the waste stream for recycling. The Act will also impose higher standards on operators of landfill sites which will raise disposal costs and therefore help to make recycling a more economically viable disposal option.

NOTES AND REFERENCES

1. HUMPHREYS, David (1988) *Mining in the UK — The Industry and its Competitiveness*, Mining Association of the UK.
2. In Scotland and Northern Ireland the registers are open for public inspection; from December 1990 on, the registries in England and Wales were also to be open to the public.
3. See report referenced in note 1.
4. Disbursements totalled £9.7 million up to 31 March 1984 and projects attracting a further £600 000 were under consideration. Since its inception in 1972, 63 companies involved in mineral exploration in 235 projects have sought assistance.

BIBLIOGRAPHY

British Geological Survey (1991), *UK Minerals Yearbook, 1990.*

COLMAN, T.B. (1990), *Exploration for metalliferous and related minerals in Britain: A Guide*, British Geological Survey.

CROWSON, P. (1977), *Nonfuel Minerals and Foreign Policy*, Royal Institute of International Affairs, London.

TOWNSHEND-ROSE, F.H.E. (1958), *Introduction to Mining Law*, Colliery Guardian, London.

US Department of Interior (1987), "The Mineral Industry of the United Kingdom", *Bureau of Mines Minerals Yearbook*, Washington, D.C.

UNITED STATES

INTRODUCTION

The US non-ferrous industry is made up of more than eleven segments with aluminium, copper, gold, lead, platinum group metals, silver, and zinc representing the major commodities produced. The country has substantial domestic reserves and occupies leading positions for the production of most of these products. In 1990, the US share in world mine production was 40 per cent for magnesium, 17 per cent for copper, 12 per cent for silicon, 14 per cent for lead, 14 per cent for silver and 7 per cent for zinc. The United States is currently the western world's largest copper refiner and second largest miner, the largest producer of aluminium and refined lead and second in lead mine output. In 1990, US zinc mine output doubled that of 1989 due to production at the Red Dog zinc mine in Alaska. The US remains one of the largest gold producers, whose production in 1990 probably outstripped that of the USSR.

Taken individually, the United States is the world's largest consuming country for all these metals. US consumption of non-ferrous metals was over US$24.5 billion in 1989. Imports to these industries were US$9.5 billion and exports were valued at US$7 billion. Canada continues to be a major trading partner and is the country with which the United States has the largest deficit in non-ferrous products. The US has a favourable balance of trade with Japan in this sector and since 1986 the country became a net exporter to the European Community.

The US share of the world market for most metals has slipped steadily over the past two decades. Throughout the 1980s a dominant theme was concern about the decline in competitiveness of the US minerals and metals industry. By late 1985, after four consecutive years of heavy losses, observers were predicting the "death of mining" in the United States and a very dim future for US metals producers.

Despite such a prediction, the US minerals and metals industry continues to compete on an international basis. The industry reached its present level of profitability after a protracted period of financial losses, restructuring, rationalisation, and capacity reduction. Factors that led to the turnaround included reduced labour costs, new applications of technology, favourable exchange rates as well as increases in commodity prices.

The United States has consistently maintained that a strong domestic minerals and metals industry is an essential contributor to the nation's economic and security interests.

Organisation of governmental agencies

Federal policy towards minerals, including non-ferrous metal mining and processing, is expressed through various laws, regulations, executive orders, treaties, rulings of commissions, and US participation in international organisations.

A major concern for the federal government is the competitiveness of the US minerals and metals industry, including balance of trade, employment and security of supply. Other interests related to the industry expressed as statutory responsibilities include the following:

- Land Use — in particular the use of federal lands on which many of the nation's ore reserves are found. Competing uses of those lands, the goals of land management, and questions of fairness in the allocation of land use are all considerations.
- Environmental Protection — as embodied in various laws regarding clean air and water and solid waste management.
- Public Health and Safety — especially in the context of occupational hazards and medical risks encountered in the work place.
- National Security — in terms of assured supply of raw materials, independent of the vagaries of politics and economics associated with imports. Especially important here are minerals and metals used in military and civilian applications essential to the national defence that are unavailable or in short supply in domestic reserves.
- Economic well-being — in terms of the health and vitality of the US minerals and metals industry.

From its inception in 1910, the Bureau of Mines (USBM) has been the principal agency of the federal government for improving productivity and safety in the minerals and metals industry. Its mission is to help ensure that the United States has an adequate and dependable supply of minerals to meet its defence and economic needs at acceptable environmental, energy, and economic costs. The US Bureau of Mines is primarily a research and statistics gathering agency. Research is conducted to provide improved technology for the extraction, processing, use and recycling of the United State's mineral resources at a reasonable cost and without harm to the environment or the workers involved. The Bureau also collects, compiles, analyses and publishes statistical and economic information on all phases of domestic and international non-fuel mineral resource development, including exploration, production, shipments, demand, stocks, prices, imports and exports. Special studies are frequently made on subjects of particular national interest, such as the effects of potential economic, technological or legal developments on resource availability.

The Bureau's mission has been statutorily enlarged in such areas as health and safety technology; conservation of the country's helium resources; research on the development, mining, preparation, treatment, and utilisation of minerals essential to the common defence or industry needs of the United States; and mineral land assessments.

In 1982, the USBM was given management responsibility for the Mineral Institute Program. The Bureau also provides the nucleus of the Federal Emergency Minerals Administration, which is chartered under the Defense Production Act of 1950, as amended, and supports the Department of Defense in its stockpiling and other materials readiness programmes.

The development of a national minerals and materials policy has been a recurring issue since World War I. In yearly 1951, when the build-up for the Korean War was hampered by shortages in metals critical to the production of war material, President Truman created the President's Commission on Materials Policy (Paley Commission). Since then the debate of minerals and materials issues has been kept alive by two commissions; several panels of the National Academy of Sciences (NAS), reports by the Congressional Research Service (CRS), the General Accounting Office and the Office of Technology Assessment (OTA).

Several advisory groups participate in the shaping of government policies with respect to minerals. The National Critical Materials Council (NCMC) advises the President on national materials policies and issues. The National Strategic Materials and Minerals Programme Advisory Committee has performed a similar function for the Secretary of the Interior. The Committee on Mining and Mineral Resources Research also advises the Secretary of the Interior on a number of matters relating to minerals research, particularly in the Mineral Institutes Programme. A fourth group, the Committee on Materials (COMAT), functions under the auspices of the Office of Science and Technology Policy in the Executive Office of the President. Through a subcommittee, the Interagency Materials Group, COMAT attempts to enhance co-operation and co-ordination between agencies involved in the support of materials research.

With a few brief exceptions, however, recommendations to bring minerals and materials policy under a single agency have never been implemented by the executive branch. In a recent National Research Council study, it was found that because the numerous national policies affecting the minerals and metals industry are not administered by a single agency, there has been no consistent policy governing minerals R&D or supply issues and no effective co-ordination of many federal activities that affect the industry[1].

PROSPECTING AND EXPLORATION

Geological services

The US Geological Survey (USGS) was established by an act of Congress in 1879 as an agency of the Department of Interior (DOI). The USGS was charged with the conducting of an "examination of the geological structure, mineral resources, and products of the national domain". Thus the agency was defined primarily as a scientific fact finding and research organisation as contrasted to a developmental or regulatory one.

The USGS is now the principal source of scientific and technical expertise in the earth sciences within the federal government. The USGS's Mineral Resources

Data System in conjunction with the USBM's Mineral Availability System provide site specific mineral deposit and occurrence data and site maps.

The USGS is organised into five divisions. The Geological Division provides geological, geochemical and geophysical information on land resources, mineral and energy resources. Its Office of Mineral Resources is responsible for assessing the non-fuel mineral resource potential of the country. The National Mapping Division provides geographic and cartographic information, maps, and technical assistance. The results of investigations are published in USGS professional papers, bulletins, circulars, and geologic and related map series.

The USGS also conducts many of its operations in collaboration with other government agencies. For example, it is heavily involved in defence mapping work with the Department of Defense and in climate and marine geology surveys with the National Oceanic and Atmospheric Administration. The USGS performs surveys of energy resources for the Department of Energy, and surveys mineral resources in conjunction with the USBM, Forest Service, National Parks Service, Fish and Wildlife Service, and Bureau of Indian Affairs.

The Office of International Geology manages and co-ordinates the Survey's International Assistance in geologic studies and scientific exchange programmes on behalf of the Department of State. The largest of the current programmes is the Mission to Saudi Arabia engaged in studies of the geology, mineral resources and natural hazards of that country. Other activities are being carried out in countries in Asia, Europe, Africa, and Central and South America on behalf of the Agency for International Development (AID) and other agencies and international institutions.

Lands availability

The access to federal lands for metallic mineral exploration is governed principally by the General Mining Law of 1872, which provides for free and open access for hard rock mineral exploration in the public domain. Explorationists obtain exclusive rights of possession to an area by staking claims and maintain these rights indefinitely, as long as they perform a minimal amount of annual assessment work.

The federal government collects no royalties from metallic minerals extracted from mining claims on its lands, but procedures are subject to the federal income tax. A claim holder can acquire free title to the surface of a mining claim if he demonstrates that an economically viable deposit exists, but purchase of a title (a patent) is not required for development or mining. Patented mining claims are private property and thus no longer federal lands.

In the last thirty-five years, even though the basic rules remain the same, this simple framework has become significantly more complicated. Exploration and mining have lost their dominant use status as three pieces of legislation (the Surface Resources Act, 1955, the Multiple Use Sustained Yield Act, 1960, and the Classification and Multiple Use Act, 1964), required a greater degree of planning and co-ordination of federal land use policies.

Federal policies designed to encourage disposal or transfer of public lands to private control have thus been replaced by policies providing for federal retention and management of many public lands. The 1964 Act reaffirmed the possibility to classify lands as off limits to mineral exploration and development.

Two pieces of legislation enacted in the middle seventies carried on and extended the multiple use philosophy of the three earlier legislative Acts. The Forest and Rangeland Renewable Resources Planning Act of 1974 governs land management of Forest Service Lands, which account for some 30 per cent of federal lands. Forest Service regulations issued in the same year require that for any activity that might cause harm to surface resources, explorationists and miners must file a notice of intent and submit an acceptable plan of operation. The regulations also require that mined land be reclaimed.

The other piece of legislation, the Federal Land Policy and Management Act of 1976 (FLPMA) also requires multiple use management on lands under the authority of the Bureau of Land Management (BLM), e.g. Indian lands, and some 40 per cent of federal lands. BLM regulations from the same period are similar to forest service regulations described above. In addition, FLPMA requires explorationists to file a record of all mining claims within 90 days of location and to submit reports of annual assessment work on claims. BLM administers mining law on both BLM and Forest Services lands.

More significantly, FLPMA authorises the Secretary of the Interior to withdraw BLM lands from mineral exploration in order to protect other resources or uses, and to review all roadless areas of 5 000 acres or more on BLM lands for possible designation as wilderness. Under the Wilderness Act of 1964, exploration is severely restricted in designated wilderness areas and areas being considered for such designation. The 1964 Act set aside some 7.7 million acres within which all development was prohibited in perpetuity. In the 26 years since the passage of the original legislation, the wilderness system has grown to nearly 36.4 million acres. Several bills have been introduced in Congress to include substantial new acreage in the system.

Other legislation under which federal lands have been withdrawn from exploration or which severely restrict such activity includes the Endangered Species Act of 1973, the Endangered American Wilderness Act of 1978 and the Alaska National Interest Lands Conservation Act of 1980. In addition, the Desert Preservation Act, if enacted into law, will effectively bar access to and development of an area of some 3.64 million acres in the desert area of Eastern and Southern California. Other laws which may impact on mining include the National Historic Preservation Act of 1966, and Amendments of 1980, and the Archaeological Resources protection Act of 1979.

Finally, superimposed on these policies is the National Environmental Policy Act of 1969 (NEPA) that requires environmental impact statements for any major federal action that significantly affects the environment. Mining and at least some types of detailed exploration are included within NEPA.

From the mining industry's perspective, two fundamental problems result from federal lands policies. The first problem is land formally closed or withdrawn from mineral exploration and development, such as most National Park lands, military bases and naval petroleum reserves. The result is less exploration in some geologically promising areas than would otherwise occur and, ultimately, higher costs of production for mining companies.

The second fundamental problem from industry's perspective is in the form of other restrictions on access to public lands remaining open for exploration, resulting in delays, added uncertainty and increased costs.

More than 30 per cent of US land is owned by the federal government, with less than 7 per cent owned by state and local governments. Federal ownership is not evenly distributed throughout the country, however. In twelve western states, including Alaska, the federal government owns some 60 per cent of all lands, whereas in the remainder of the country it owns less than 5 per cent of the land. Over 90 per cent of federal lands are in these twelve states, with more than 40 per cent in Alaska alone.

Taxation and other incentives

Minerals producers may deduct a maximum of 70 per cent of the cost of exploration and development in the year incurred, and capitalise the remaining 30 per cent over a five year straight line depreciation schedule. This defers a portion of income taxes until the deductions have been taken. For development expenditures, only the amount that exceeds the net receipts of the mine for a given year may be included.

The Defense Production Act of 1950 authorised and financed two programmes whereby the government and private industry jointly participated in minerals exploration projects. Under this project, 1 159 contracts were executed to explore for strategic and critical minerals essential to national defence.

A successor programme, USGS's Office of Minerals Exploration was established in 1958 for the purpose of exploring for minerals needed in an expanding economy in joint ventures with private industry. The programme operated through 1974 when lending authority was terminated. It provided government financial assistance of an average of 62 per cent of the allowable costs of exploration for a number of substances up to a limit of US$250 000. Repayment to the government was at the rate of a 5 per cent royalty on production from the property. There was no requirement to produce, and if there was no production, no repayment was required.

MINING AND METALS PRODUCTION

Taxation

Mining and metal companies are subject to federal and state income taxes as well as sales, social security, capital, and severance taxes. Of all these taxes, national income taxes probably have the greatest influence on mineral industry competitiveness. The Tax Reform Act of 1986 reduced the top corporate tax rate from 46 per cent to 34 per cent, set the minimum tax at 20 per cent, and changed the depreciation schedule for mining equipment from five years at 150 per cent to seven years at 200 per cent. Most states impose state income and/or franchise taxes at rates varying from a nominal filing fee to about 12 per cent.

Exploration expenditures may be deducted as incurred. However they are recaptured if the property is developed and attains commercial production. The deductions are recaptured either immediately or as income is realised. Development expenditures may be either deducted as incurred or deferred and amortised as the ores benefited by these expenditures are sold.

Existing depletion allowances enable mineral producers to deduct a percentage of taxable net income based on either investment cost (cost depletion) or a specified fraction of gross sales from the minerals extracted (percentage depletion) whichever is higher. The applicable percentage depletion rates vary from 15 per cent in the case of copper and silver to 22 per cent for nickel, lead and zinc. In recent years, the depletion allowance has been limited to 50 per cent of taxable net income.

Net operating losses and unused investment tax credits may be carried back three years and forward fifteen.

Direct and indirect support

The US Government seldom intervenes in the financial affairs of domestic industries that operate within the laws. Rarely has financial aid been granted to an industry suffering grave financial distress unless it is believed that the failure of the industry is a threat to national security. In recent history there have been no examples of this kind of assistance for the minerals and metals industry.

There is no government ownership or direct control over business enterprises in the minerals and metals industry.

HORIZONTAL POLICIES

Foreign investment controls

The United States imposes few restrictions on foreign investment. US subsidiaries of foreign companies are incorporated under US laws, and are generally granted the same privileges and assume the same responsibilities as any US corporation.

The Mining Law of 1872 permits ownership of leases and claims on federally administered lands by all US citizens and corporations, regardless of the nationality of the stockholders, with one exception. Under the Minerals Leasing Act of 1920, if the majority of the stockholders of a US corporation are of a country that does not allow equal access to US investors in the foreign country they represent (i.e. if reciprocal investment agreements do not exist), then the foreign-held US subsidiary may be denied access to mineral rights on federally administered land. However, the reciprocity laws do not exclude foreigners from exploration, purchase, or development of private land, regardless of whether the investment is through a US subsidiary or directly by a foreign operation.

While US policy rarely distinguishes between domestic and foreign investors, the federal government does monitor foreign investment. A provision of the Omnibus Trade and Competitiveness Act of 1988 authorised the President to prohibit any foreign investment transaction that would threaten to impair national security. The President delegated his authority to review foreign take-overs to the cabinet level committee on foreign investment in the United States. This Committee investigates foreign transactions on a case-by-case basis. If conditions warrant, the Committee may recommend that the President prevent individual foreign take-overs or mergers that threaten to impair national security. Of the more than 500 cases reviewed by the Committee on Foreign Investment in the United States (CFIUS) since 1988, the President has moved to block only one investment; the proposed take-over of an aerospace company by a firm closely related to the Government of China.

Environment

The minerals and metals industry is subject to numerous federal and state regulatory requirements related to environmental protection. These range from the preparation of environmental assessment and/or an environmental impact assessment prior to initiating a mining project if federal approval is needed, to the control of air and water pollution during mining and processing, to the reclamation of the tailing piles and dumps when an operation closes.

Mineral operations are also scrutinised by the Mine Safety and Health Administration and the Occupational Health and Safety Administration. As previously mentioned, other types of legislation either regulate the location of mines on public lands or withdraw those lands from mining altogether.

The landmark legislation is the National Environmental Protection Act (NEPA), approved by Congress in 1969, that called for the formation of the Environmental Protection Agency (EPA), a Council on Environmental Quality (CEQ) to advise the President on environmental matters, and laid the basis for regulations which required the preparation of an Environmental Impact Statement (EIS) for all major projects sponsored by the government, including mine development. Environmental impact statements are not required by private individuals operating on private lands.

An EIS must be prepared and circulated before government approval of any action can be taken which significantly affects the quality of the human environment. The Forest Service may employ a still less stringent Environmental Analysis. Each agency has its implementation regulations patterned after the guidelines published by the Council on Environmental Quality.

The decades of the seventies and the eighties saw a proliferation of environmental control legislation which impose regulations on day-to-day operating practices in the mining industry, such as the Clean Air Act (1970), the Clean Water Act (1972), the Resource Conservation and Recovery Act (1976), known as RCRA, and the Comprehensive Environmental Response, Compensation and Liability Act (CERCLA) (1980), commonly referred to as the Superfund Act.

In addition to these federal statutes, most states have enacted a similar, though often more stringent, set of regulations pertaining to mining and processing operations. In many cases, the state regulations are tied into their federal counterparts, thus allowing individual states to take responsibility for implementation and enforcement of the federal regulations.

By federal legislation, the EPA is a purely regulatory agency with a leading responsibility for overseeing environmental protection activities over air, water, the disposal of toxic wastes, and Superfund clean-up activities. The EPA is organised into ten geographic regions with a high degree of local autonomy. Regions tend to be organised into the general regulatory classifications of air, water, toxics and wastes, and Superfund.

In parallel, regulatory agencies at both the federal and state levels have proliferated. These agencies perform a variety of functions including the issuance of permits, monitoring of operations, regulatory enforcement and verification of compliance. EPA is encouraged to delegate many of its responsibilities to those states who wish to and are capable of taking over those duties.

A recent study by the Congressional Research Service[2] has estimated that the cost of complying with federal environmental regulations is about 6 cents per pound of lead and between 9 and 15 cents per pound of copper — about 20 per cent of the price of each metal in 1986, though rising metal prices have reduced this fraction to about 10 per cent in 1990.

Clean Air Act

The Clean Air Act sets standards for both ambient concentrations of pollutants and emissions from individual sources. It requires each state to develop an approvable State Implementation Plan to correct air pollution problems in the state. The National Ambient Air Quality Standards (NAAQS), which address ambient concentrations, include primary standards designed to protect human health and secondary standards to safeguard public welfare.

The EPA has set primary and secondary standards for sulphur oxides, particulate matter, nitrogen dioxide, hydrocarbons, photochemical oxidants, carbon monoxide, ozone, and lead.

Every major new source of emissions (e.g. a new smelter furnace) is required to undergo a pre-construction review to ensure it will not violate NAAQS. Sources in dirty air areas, or at the opposite extreme, those where the air is already much cleaner than the standards require, are subject to more stringent permitting requirements for new sources. In addition, operating sources are required to use technological controls to meet emission limitations, which set quantitative limits on the amount of pollutants that can be released to the atmosphere.

Clean Water Act

The Clean Water Act establishes water quality standards that focus on the uses of the waters involved, including public water supplies, fish and wildlife, recreation, and agriculture. The standards generally are achieved through effluent limitations that restrict the quantities, rates, and concentrations of chemical, physical, biological, and other types of discharges from individual sources. In general, the Act requires all categories of sources (including mines, mills, smelters and refineries) to apply the best practicable control technology currently available in order to meet the effluent limitations.

Effluent limitations and water quality standards are implemented through State certification programmes and the National Pollutant Discharge Elimination System (NPDES). All point sources must obtain state certification that their operations will not violate any effluent limitations, water quality standards, or new source performance standards. They also must obtain a NPDES permit, which requires a demonstration that the discharge will meet all applicable water quality requirements. NPDES permits are issued under EPA approved State programmes.

Effluent limitations for mines, mills, and leach operations cover discharges for copper, zinc, lead and cadmium, as well as total suspended solids and pH. Arsenic and nickel are not specifically mentioned in the standards because they are adequately controlled by the removal of other metals found in the discharges. Leaching operations generally are expected to achieve zero discharge unless the annual precipitation exceeds annual evaporation (rare in the arid and semi-arid copper producing areas of the Western United States).

Guidelines are being developed for effluents discharged from smelters, refineries and acid plants. These limitations aim to control the amount of arsenic, cadmium, copper, lead, zinc and nickel in effluents; the pH of the discharge; and the concentration of total suspended solids.

Safe Drinking Water Act

Congress enacted the Safe Drinking Water Act in 1974 to ensure that water from public drinking supplies is healthful. It provides for a Public Water Systems Supervision Program with enforceable Maximum Contaminant Levels (MCLs) for drinking water standards. Primary standards are set based on the contaminant concentrations at which no known or anticipated adverse effects on human health occur, modified by the best available treatment technology (considering cost). Secondary standards set goals for contaminants that primarily affect the aesthetic qualities of drinking water.

The Safe Drinking Water Act also protects aquiferous sources (those that supply 50 per cent or more of the drinking water for an area) from contamination due to projects above the aquifer. It requires states to establish "well head protection areas" around public wells to prevent pollutants from entering underground supplies.

Resource Conservation and Recovery Act (RCRA)

The EPA regulates hazardous and other solid wastes under the Resource Conservation and Recovery Act (RCRA). Subtitle C of RCRA establishes regulations for the generation, transportation, treatment, storage and disposal of materials identified by EPA as hazardous. Subtitle D provides Federal guidelines for EPA approved state or regional solid waste plans. These address the regulation of landfills, dumps and ponds handling non-hazardous solid and liquid wastes.

Hitherto, mining wastes had been excluded or exempt of RCRA by the so called Bevill Amendment. In July 1986, an EPA decision determined that solid wastes from the mining and beneficiation of mineral ores should be regulated under Subtitle D of RCRA rather than as hazardous wastes under Subtitle C. The rationale for this decision was that the large volumes of mine waste would be very difficult to regulate under rules that had been designed to manage much smaller amounts of hazardous industrial and municipal waste. EPA also reasoned that Subtitle C does not allow considerations of environmental necessity, technological feasibility and economic practicality, which are important given the magnitude of mine waste.

In 1989, a court order directed EPA to narrow the Bevill exemption by specifically defining the mineral process wastes in terms of low toxicity and high volumes. EPA then identified a total of 18 waste streams deemed to be exempt from the Bevill Amendment and therefore subject to regulation. EPA's Office of Solid Waste (OSW) is currently preparing a programme based on the authority of Subtitle D to regulate the mentioned 18 exempt wastes and others not listed as being potentially hazardous. As a point of departure, OSW issued in 1988 a Strawman, or working document, for the purpose of eliciting comments by interested parties. A second Strawman was released in 1990. The final mineral processing regulatory determination was made in 1991.

Comprehensive Environmental Response, Compensation and Liability Act (CERCLA) or Superfund

Superfund allows the federal or state governments and in some cases, private parties, to assess and clean-up the release of hazardous substances into the environment. It also allows a response at the mere threat of a release of hazardous substances. A fund of more than US$9 billion was established to encourage clean-ups. The law was enacted because of the public concern about abandoned, hazardous waste sites, but it also requires response actions to be taken wherever a release or threatened release has or may occur. Thus, it applies to many facilities.

EPA is authorised to spend Superfund money, to issue administrative orders, to go to Court to get compliance with its assessment and clean up plans, and to recover all of its costs from the parties responsible for the release. Responsible parties are considered to be owners of the contaminated land, operators of facilities that release the hazardous substance, transporters and the original generator of the material.

CERCLA applies to closed and abandoned mining facilities where hazardous substances were or are being released. Most metals are listed as potentially hazardous substances, so EPA has CERCLA authority to clean them up. Under the recent RCRA corrective action proposal rule, EPA also will have authority for the clean up of hazardous waste at offsite facilities.

Although mining wastes are exempt from RCRA Subtitle C regulation, EPA has made it clear that such materials are not exempt from Superfund. The EPA's policy on the continuing availability of the mining waste exclusion for inactive or closed facilities will affect the extent to which Superfund liabilities and obligations may arise from the closure of a facility. Therefore, when considering closure, the potential application of immediate or future hazardous waste regulatory scrutiny must be evaluated.

Occupational Safety and Health Administration (OSHA)

The Occupational Safety and Health Act of 1970 was passed to assure so far as possible every working man and woman safe and healthful working conditions. The Act charges the Secretary of Labour with promulgating and enforcing safety and health standards to meet this objective. OSHA is pre-empted from enforcing its standards when another federal agency exercises its statutory authority for workplace safety and health conditions. The purpose of this provision is to provide comprehensive protection to all employees while avoiding duplication of employee safety and health regulations by OSHA and other federal regulatory agencies.

In a 1980 Supreme Court ruling[3], however, the Occupational Safety and Health Administration (OSHA) was required to find the existence of a significant safety and health risk under a current standard before issuing a new regulation and to demonstrate that a new standard would reduce or eliminate the risk. Specifically, the Court stated that "before he can promulgate any permanent health or safety standard, the Secretary [of Labour] is required to make a threshold finding that a place of employment is unsafe — in the sense that significant risks are present and can be eliminated or lessened by a change in practices".

Mine Safety and Health Administration (MSHA)

Under the Federal Mine Safety and Health Act (Mine Act), the Mine Safety and Health Administration (MSHA) administers programmes to improve safety and health conditions in US mines. MSHA's objectives are achieved by (i) developing and enforcing mandatory safety and health standards; (ii) providing technical assistance to mine operators and approving certain equipment for use in underground mines; and (iii) encouraging and expanding education and training programmes.

To achieve its regulatory objectives, MSHA has devoted considerable resources toward a comprehensive review of the Agency's existing body of technical and work practice standards. This review is designed to delete irrelevant, duplicative, and outdated standards; eliminate conflicting provisions, minimise reporting and record keeping burdens; include compliance alternatives, when possible, upgrade requirements consistent with advances in mining technology, include performance-oriented standards; and address known hazards not now covered by regulations.

In developing safety and health standards, the Agency is required to be cognizant that the Mine Act prohibits MSHA from promulgating any standard that reduces protection to miners under existing standards.

Effects on the non-ferrous metals industry

In the last twenty years environmental regulations have proved to be a heavy burden for the US metals industry. In fact, environmental constraints, contributed to the permanent closure of five smelters and three copper refiners in the eighties. Depletion of domestic reserves coupled with environmental costs led to a dramatic reduction in US slab zinc capacity, from 1 million tonnes in 1970 to 325 000 tonnes in 1989.

Acid rain legislation, because of its effects on electric utilities, is the principal environmental concern of the aluminium industry, as proposed SO_2 emissions regulations will raise costs at coal-based utilities. However, the most important environmental issue for the minerals industry concerns the health effects of lead. Health concerns have eliminated demand for the metal in some sectors (e.g. gasoline additives, indoor paints and various building products) and there is also concern regarding the health and safety of workers exposed to lead particulates during the production process. Most of the US industry is not in compliance with the current National Ambient Air Quality Standards (NAAQS) particulate standard for lead[4].

Research and Development (R&D)

Several agencies of the federal government provide support for mining and minerals research and technology development. They are the:

Department of the Interior (DOI)

The leading Agency, accounting for the great majority of federal research funding in the field, is the Department of the Interior (DOI) through its Bureau of Mines (USBM). Bureau support of R&D relevant to the needs of the industry contributed substantially to its growth and competitiveness throughout much of this century, particularly in the early decades. However, the Bureau's technological contributions have had less impact in recent years. Its R&D budget was lower in 1989 (even before inflation is taken into account) than it was in 1980.

The USBM research programmes focus variously on improvements in exploration and mining technology; minerals and materials science and processing technology; health, safety, and environmental technology; and ancillary programmes such as methods for improving process management and management technology (e.g. computer control). The nature of the research ranges from fundamental to highly applied, although the emphasis is strongly toward the applied end. Some 40 to 50 per cent of the research is devoted to mine safety, health and environmental protection, further limiting the funds available for technology to improve productivity[5].

The USBM research is conducted both in-house and at university laboratories. Under its Mineral Institutes Program, USBM sponsors a number of State Mining and Mineral Resources Research Institutes and Generic Mineral Technology Centers (GMTCs). Currently there are 30 mineral institutes that function as administrative mechanisms for the distribution of funds to academic departments for research in the mineral sciences and engineering. The overall budget of the programme was US$12.5 million in 1990.

As part of the Minerals Institute Program, research grants are also made to six GMTCs covering major aspects of the minerals industry. The GMTCs are located at universities with mineral institutes and are intended to facilitate government-industry-university co-operation in generic areas.

The US Geological Survey (USGS) also maintains programmes of research under its Office of Mineral Resources. This research involves the theoretical and technical aspects of mineral exploration and resource assessment. It encompasses both geochemical and geophysical methods for locating and modelling mineral deposits in relation to the environments in which they occur. Other projects support the preparation of resource maps and information systems that are valuable exploration tools for the industry. This research is carried out through field studies as well as laboratory experimentation.

The USBM and USGS research efforts are complementary in support of exploration and economic assessment of mineral deposits.

National Science Foundation (NSF)

Another agency providing support for minerals research is the National Science Foundation (NSF). Through its Materials Engineering and Processing Program, the NSF funds a small amount of work in extraction, smelting, and solidification processes. The Division of Materials Research supports a programme of research focused on basic scientific research in physical metallurgy. NSF research in related topics, such as tunnelling technology, are also of benefit to the mining industry. It also supports research in the geosciences.

Department of Energy (DOE)

The Department of Energy (DOE) has more than 20 programme offices responsible for aspects of materials research from basic to applied. DOE has the largest single materials research programme in the federal government but much of its work is directed at materials science for advanced energy related materials, processing and refining electronic materials and specialised alloys, uranium and weapons grade plutonium. The DOE has prepared a research programme that identifies specific opportunities for research to contribute to the competitiveness of the aluminium and copper industries through increased energy efficiency.

Department of Commerce (DOC)

The Department of Commerce (DOC) supports mining and minerals research through two of its agencies, the National Institute of Standards and Technology (NIST) and the National Oceanic and Atmospheric Administration (NOAA). NIST currently supports a single project covering the bio-active extraction of metals. Some of its work on sensors for use in processing is relevant to metals processing. NOAA, through its Sea Grant programme, sponsors a limited amount of academic research related to ocean mining and minerals.

Others

At the Department of Defense (DOD), the Defense Advanced Research Projects Agency, the Strategic Defense Initiative Office, and the Military Services (particularly the US Air Force) conduct research in advanced materials processing and manufacturing, including physical metallurgy, extrusion, rolling and joining. The National Aeronautics and Space Administration (NASA) sponsors research in many of the same areas.

Other federal agencies do not support significant amounts of R&D in this area, but provide input into the research programmes of the USBM and others. These contributing agencies include the Mine Safety and Health Administration, the National Institute for Occupational Safety and Health, the Office of Surface Mining Reclamation and Enforcement, and the Environmental Protection Agency.

Federal expenditures for mining and minerals-related R&D in 1989 are shown in Table V.

Table V. **United States: federal expenditures for mining and minerals-related R&D in 1989 (million US$)**

Agency/Programme		Federal expenditures
Bureau of Mines		100.9
Health,Safety and Mining Technology	51.7	
Minerals and Materials Science	24.6	
Environmental Technology	14.6	
Minerals Institutes and Generic Centers	10.0	
US Geological Survey		23.0
Development of Assessment Techniques	10.0	
Strategic and Critical Minerals	3.7	
National Mineral Resource Assessment	9.3	
National Science Foundation		0.7
Department of Energy		12.2
National Institute of Standards and Technology		0.3
National Oceanic and Atmospheric Administration		1.3
TOTAL		138.4

Trade Policy

General

The International Trade Administration (ITA) of the Department of Commerce and the Office of the United States Trade Representative (USTR), the trade negotiating wing of the Executive Branch, work jointly on most trade-related matters affecting the non-ferrous metals industry. ITA and USTR, in consultation with the private sector and other federal agencies, represent the US in most multilateral and bilateral trade negotiations directly affecting the non-ferrous metals industry. ITA and USTR receive private sector advise from the Industry Sector Advisory Committee for the Non-ferrous Metals Industry (ISAC-11). ISAC-11 was established under the Trade Act of 1974 as a two-way line of communication between federal trade officials and non-ferrous metals industry executives.

The Bureau of the Census of the Department of Commerce publishes monthly trade data, as well as annual data on shipments, employment, new capital expenditures and inventories of the non-ferrous metals industry.

The International Trade Commission (ITC) is an independent, quasi-judicial government agency established by Congress with broad powers to investigate all factors relating to the effect of US foreign trade on domestic production, employment and consumption. The ITC makes determinations as to whether imports injure or threaten injury to US industry, including cases where imports are said to be sold at less than fair value or are subsidised by foreign governments. A finding by the ITC can result in quantitative restrictions or extra duties being applied by the Department of Commerce to certain imported products to provide relief for domestic industries. The ITC is not charged with a policy-making role, but contributes to policy development by providing fact-finding reports and advice to the President and Congress concerning proposed trade agreements and trade legislation.

The United States Government advocates free and open trade with regard to non-ferrous metals. Ores and concentrates of aluminium, nickel and tin enter the US duty free. The tariff on lead ores and concentrates and zinc ores and concentrates are 1.7 cents/kg and 0.7 cents/kg respectively. The lead content of copper ores and concentrates are also subject to a tariff of 1.7 cents/kg. Tariffs on semis and finished products are generally below 6 per cent in the case of all six metals, with certain exceptions. Tariff peaks exist on tube or pipe fittings of refined copper (11.2 per cent), powders and flakes of lead (11.25 per cent) and casting grade zinc and zinc alloys (both 19 per cent).

The US has enacted several different types of legislation aimed at protecting US producers and manufacturers from the unfair trade practices of foreign competitors. Section 201 of the Trade Act of 1974 (19 USC 2251) provides for temporary relief from imports for domestic industry when imports are in such increased quantities as to be a substantial cause of serious injury or create the threat of serious injury to a domestic industry. Import relief may be in the form of a tariff, tariff-rate quota or quantitative restriction (quota). Under the law, the President is to provide relief unless he finds that the costs to the nation of providing such relief outweigh the benefits.

In the case of the minerals industry, one can point to the case of the copper industry that asked the ITC, pursuant to section 201 of the Trade Act of 1974, for protection from unfair competition from overseas producers in 1977 and again in 1983. In both cases, although the ITC found that imports of copper were a substantial cause of serious injury to the US copper industry, the President determined that it was not in the national economic interest of the United States to impose any trade remedies. In other cases during 1986-1988, pursuant to Title VII of the Tariff Act of 1930, the ITC found that the US industry was being injured by imports of certain brass sheet and strip products from various countries. Because the Department of Commerce found that these products were being sold at less than fair value, imports were subjected to additional duties.

In the case of dumping or subsidisation, the ITC conducts an investigation to determine whether or not US industry is materially injured, is threatened with material injury or if the establishment of a US industry is materially retarded due to imports. If, after an investigation, the ITC makes an affirmative injury determination, the case if referred to ITA. ITA determines whether dumping or unfair subsidisation has taken place, and calculates a dumping margin or subsidy rate. US Customs is directed to collect additional duties equal to the dumping margin or subsidy rate on imports of the product in question. One year after an anti-dumping or countervailing duty order is imposed, an interested party can request ITA to conduct an administrative review. The purpose of the review is to establish the actual dumping margin or subsidy rate for a given review period. The duties paid by the importer will be adjusted to reflect the actual duty calculated. There are currently fifteen anti-dumping or countervailing duty orders in force on non-ferrous metals products. Most of these orders cover brass or aluminium products.

Import and export restrictions

The United States regulates the export of some raw materials and products under the Omnibus Trade and Competitiveness Act of 1988, the Export Administration Act of 1979 (as amended by Executive Order 12730), the Export Administration Act of 1969 (as amended by Executive Order 12002), and the Arms Export Control Act (22 USC.2778E). The United States has no export restrictions on copper, lead, or tin minerals or metals. There are export restrictions on certain zinc powders used in missile propellants, certain nickel powders used in gas diffusion separation of nuclear fuel and certain other nickel- and aluminium-based products used in the production of aircraft turbines and other aircraft parts.

The United States currently bans the importation of all products from Cuba, North Korea and Vietnam under the Trading with the Enemy Act (50 USC Appendix 5). Nickel from Cuba and zinc from North Korea are notable examples of non-ferrous metals affected by this ban.

Government procurement

The Buy American Act (41 USC 10a *et seq.*) was enacted to give preference to domestically-produced articles, materials or supplies purchased for use by the federal government, except in cases where: (i) US goods are not available in sufficient quantity and satisfactory quality; (ii) the cost is unreasonable or; (iii) it is inconsistent with the public interest to purchase US articles, materials or supplies. Executive orders have specified that the cost is unreasonable if the US price exceeds the foreign price by more than 6 per cent, (12 per cent if a US small business price and 50 per cent if a Department of Defence procurement). Examples of non-ferrous minerals and metals falling under the first exception are bauxite, chrome ore or chromite, nickel (primary nickel, nickel ingots, cathodes or similar forms and nickel oxide and salts), tin in bars, blocks and pigs and tungsten. The Buy America Act applies strictly to United States Government procurement.

The United States is a signatory to the GATT Government Procurement Code, which was negotiated during the Tokyo Round. Implementing legislation for the Code is found in the Trade Agreements Act of 1979. As a signatory to the Procurement Code, the United States has an obligation not to apply the Buy American Act preference for domestic products or impose other discriminatory procurement measures on Code-covered purchases when designated suppliers from Code signatories or other designated countries are bidding on a contract. The United States does not apply the Code to purchases made under small business set aside programmes and to national security purchases.

The United States-Canada Free Trade Agreement

The United States and Canada implemented an accord in January 1989 that seeks to liberalise trade and investment between the two countries by establishing a free trade area. This bilateral agreement eliminates all tariffs on goods trade by 1998, reduces non-tariff trade barriers, establishes rules for bilateral investment, and creates a dispute settlement mechanism. The United States-Canada Free Trade Agreement (FTA) was approved by the US Congress and the Canadian Parliament.

The FTA was opposed by the Non-Ferrous Metals Producers Committee (NFMPC). The NFMPC alleged that Canadian minerals companies were using below-the-market-rate capital from various national and provincial government assistance programmes to modernise facilities. The legality of such programmes is to be resolved by a bilateral working group established to work out the differences between American and Canadian unfair trade law. Until the group finishes its work (up to seven years), both countries will apply their own anti-dumping and countervailing duty laws to any disputes that may arise. For cases under these laws that are investigated during this interim period, the FTA comes into play at the end of the proceedings, after the ITC and the Commerce Department (or their Canadian counterparts) have made their final determinations. Independent bi-national panels will review contested determinations for their consistency with the laws of the country that made them. Prior to the FTA's implementation, national courts exclusively undertook such reviews.

North American Free Trade Agreement

On 12 August 1992, the United States, Canada and Mexico reached an agreement in principle on a North American Free Trade Agreement (NAFTA). The Agreement is expected to be implemented in the three countries' respective domestic legislations by 1994. A NAFTA will promote the efficient use of minerals and metals in the North American region as a whole, encouraging economies of scale and promoting intra-industry specialisation.

CONSUMPTION/SUPPLY

Dependence of the United States on a few nations of uncertain reliability for materials that are essential to many industrial and defence uses has, since World War II, heightened concern over materials and minerals policy and US vulnerability to interruptions of supplies of strategic materials.

There is no single generic approach to reduce materials import vulnerability; to be effective, different actions must be taken for each metal under consideration. The overall strategy to reduce US reliance on uncertain sources of supply of strategic materials is based on a combination of three technical approaches:

— increase the diversity of world supply of strategic metals through exploration and development of promising deposits, both foreign and domestic;
— decrease demand for strategic metals through the implementation of improved manufacturing process and recycling of strategic materials from scrap and waste;
— identify and test substitute materials for current applications and develop new materials with reduced strategic material content for future applications.

Government actions to assure secure supplies of metals critical to the United States have been limited largely to reliance on the national defence stockpile to ensure the availability of materials required for national defence in time of war, leaving it to the free market to provide a diversity of suppliers for the industrial economy.

In the long term, federal funding of strategic materials R&D in the areas of recycling, substitution and advanced materials can provide more secure sources of supply that reduce the need for strategic materials.

Securing overseas supply

Minerals exploration in the United States began in earnest about 1850, leading to the discovery of some of the world's richest deposits of many major minerals. By the latter decades of the nineteenth century, however, growth in world-wide demand began to stimulate interest in global exploration. By the turn of the century, base metals in the Americas were being mined in Canada, Australia and Mexico. At the beginning of the World War I, several large copper projects were being worked by US companies in Chile and Central Africa and rich lead, zinc and silver ores were flowing from Peru. World War II greatly accelerated the depletion of higher grade ore reserves in the United States. New deposits were discovered and new mines and plants were built throughout the world. US firms owned, controlled or exploited many of these new foreign deposits.

US minerals firms' success abroad and the continued importance of the domestic mining industry help explain why official support to exploration abroad has not been as prominent as in some other OECD countries. Nonetheless the system of tax benefits of depletion allowances may also be used to reduce tax liabilities on overseas income, but rates are lower than those applying for domestic activities. Thus, depletion allowance rates for bauxite, copper, lead, nickel, tin and zinc are all 14 per cent, instead of the 15 and 22 per cent rates applicable for domestic activity.

Several US Government agencies contribute to improve the conditions of operation of US companies in cases of uncertain legal environments or restrictive foreign investment laws. They include the Department of State, the Department of Commerce and the Export-Import Bank. US participation in international development banks and United Nations activities also contribute to improve both the political and economic climate for the development of strategic materials in specific countries.

The Overseas Private Investment Corporation (OPIC) is a self-sustaining US Government agency whose purpose is to encourage private investment in developing countries. OPIC assists investors in this effort through two major programmes, the insurance of investments against certain political risks and the financing of investments through direct loans and/or loan guarantees. All insurance guarantees are backed by the US Government as well as the reserves of OPIC.

For investments in eligible countries, OPIC insures against three kinds of risk:

i) inability to convert into dollars local currency received by the investor as profits or earnings or return of the original investment;

ii) loss of investment due to expropriation, nationalisation or confiscation by action of a foreign government;

iii) loss on income and loss of assets due to political violence, i.e. war, revolution, insurrection, or civil strife.

OPIC finances overseas ventures wholly-owned by US companies and joint ventures between local companies and US sponsor firms. The US firm must have at least a 25 per cent equity interest in the project. In addition, any OPIC guaranteed loan must be approved by the host country where the project borrower is located. Additionally, projects must not have an adverse effect on the US economy or employment situation.

For minerals projects OPIC may provide:

i) coverage for up to 90 per cent of the initial investment plus an equal amount of retained earnings. The percentage of coverage depends on OPIC's assessment of risk factors in the project and multinational equity or debt participation;

ii) insurance against all three covered risks during the exploration phase for intangible costs as well as tangible assets;

iii) special additional coverage against losses resulting from the breach of certain host government undertakings identified by the project sponsor at the beginning as being vital to the success of the project;

iv) in addition to standard political violence coverage, insurance to cover consequential loss due to the closing of operations for a period of at least six months because of political violence in the project country or by certain specified political violence in another country.

Stockpiles

The most visible policy taken by the United States to guard against disruptions of supplies of strategic materials is the National Defense Stockpile. A strategic material may be briefly defined as one for which the quantity required for essential civilian and military uses exceeds the reasonable secure domestic and foreign supplies, and for which acceptable substitutes are not available within a reasonable period of time.

The objective of the stockpile is to support the US defence, industrial and essential civilian requirements during a prolonged military conflict or declared national emergency. As the contents of the stockpile are reserved for national security purposes only, the US industrial economy is vulnerable to a variety of supply disruptions that may arise in times of peace.

First established in 1939 through the Strategic and Critical Materials Stock Piling Act of 1939, the stockpile was built up rapidly in the post-World War II and Korean War era, when the goal of the stockpile was to accumulate a five-year supply of critical materials. Many of the materials now in the stockpile date from this period, and may be antiquated due to changes in the materials specifications [6]. During the Vietnam War period, substantial amounts of stockpile material were declared excess to stockpile needs and sold by successive administrations, a circumstance that lead to charges that the stockpile was being used to keep metal prices stable during the Vietnam War.

Stockpile goals had been reduced from the initial five years to three years in 1958, and to one year in 1973, before being raised back to three years in 1976 during the Ford Administration. In 1979, the Strategic and Critical Material Stockpiling Revision Act reaffirmed the need for stockpiling, as well as conservation and development of domestic sources of strategic materials.

The Act specifies that "the purpose of the stockpile is to serve the interest of national defense only and is not to be used for economic and budgetary purposes" and that the "quantities of material stockpiled should be sufficient to sustain the United States for a period of not less than three years in the event of a national emergency".

Stockpile goals are based on having a three year supply of materials needed to meet national defence and industrial needs in a defence emergency. A transaction fund dedicates revenue from federal sales of stockpile excesses to the purchase of materials short of stockpile goals. In September, 1990, the total stockpile inventory was valued at approximately US$9.4 billion.

Mechanisms for assuring availability of materials and industrial capacity needed for national security include loans or loan guarantees for corporate activities that would expedite production in the event of a national emergency. These include expansion of capacity, development of technological processes, or the production of essential materials, including exploration, development and mining of strategic metals. Under the Defense Production Act (DPA) of 1950, the government may also purchase metals and minerals for government use or resale.

In 1981 the Reagan Administration announced a major new stockpile acquisition programme, aimed at meeting stockpile goals for 15 priority materials. In 1984, a programme was launched to upgrade chromite ore and manganese ore to high carbon ferrochromium and high carbon ferromanganese. This project was initiated to help sustain a US ferro-alloy furnace and processing capability vital to the National Defense.

In the 1984 reauthorisation of DPA, Congress established new procedures for authorisation of the previously mentioned projects in the absence of a national emergency or war. The law requires the President to determine that federally supported projects meet essential defence needs and that the federal support offered would be the most cost-effective, expedient, and practical alternatives for meeting the need.

Although DPA provisions generally have been used to encourage mining of strategic minerals, the law also could be used to ensure adequate smelting and refining capacity to meet domestic security needs, and to develop advanced technologies considered desirable for enhancing the security of US domestic resources.

The National Defense Authorization Act of 1987 reaffirmed the stockpile goals, but authorised the President to make changes of less than 10 per cent after submitting a justification to Congress and waiting until the beginning of the next fiscal year after submission. Changes greater than 10 per cent require express authorisation by law.

The US Secretary of Defense is the stockpile manager and the procurement, storage, rotation, and disposal activities are managed by the Defense Logistics Agency (DLA) of the Department of Defense. Data on world-wide supplies and requirements provided by the Department of Commerce and the Bureau of Mines to the Institute of Defense Analyses, which maintains computerised models to assist the Department of Defense in making necessary determinations.

Recycling

The federal government influences recycling both directly end indirectly. Direct effects arise from its R&D activities aimed at improving recycling technologies, namely for strategic materials. Indirect effects, some positive and some negative, arise from a wide variety of federal policies and programmes that are not specifically aimed at recycling. Such activities as taxation, regulation of transportation rates, environmental policy, and procurement and disposal of materials by federal agencies affect material flows in the economy.

The National Materials and Minerals Policy, Research and Development Act of 1980 was intended to provide a basic co-ordinating framework for executive branch materials policy decisions. The Act encompasses all materials related to industrial, military, and essential civilian needs. It emphasises, however, strategic materials for which the United States is heavily import dependent but could augment supply through substitution, recycling, and conservation.

Environmental factors caused non-ferrous metals recycling to become more difficult in the 1980s. Lead's health effects are the industry's most important environmental issue.

Lead recycling in the US increased over the last few years, reflecting the combined efforts of government and the lead producing and using industries to reduce or eliminate dissipative uses of lead. Spent lead batteries are classified by the EPA as hazardous waste under the Resource Conservation and Recovery Act, but have a special exemption for collection, handling and storing until delivery at the smelter. Many scrap collectors and dealers refuse to handle batteries for fear of incurring liability under Superfund. Almost half of the states have enacted laws that mandate the recycling of automotive batteries, while the EPA continued efforts to establish a regulatory framework to prevent lead pollution.

The concepts developed for lead could be applied to other US minerals and metals industries by the year 2000. Other proposals include lowering exposure and emission standards and tightening the regulation of hazardous materials and wastes.

NOTES AND REFERENCES

1. National Research Council, 1990.
2. Congressional Research Service (1986), *Competitiveness of American Metal Mining and Processing*, report prepared for US House of Representatives Committee on Energy and Commerce, Sub-Committee on Oversight and Investigation, Committee Print 99-FF, Washington, D.C.
3. Industrial Union Department, AFL-CIO v. American Petroleum Institute.
4. Office of Technology Assessment, 1990.
5. National Research Council, 1990.
6. Office of Technology Assessment, 1985.

BIBLIOGRAPHY

Bureau of Mines (1990), *State Mineral Summaries*, Washington D.C.

Bureau of Mines, *Transparency of Markets Study, US Non-ferrous Metals Industry*, internal paper, Division of Policy Analysis.

Department of Defense (1990), *Strategic and Critical Materials Report to the Congress*, Washington D.C.

DEYOUNG Jr. J.H. (1991), *Transparency of Markets: US Mining and Metals Industry*, internal paper.

EGGERT, R.G. (1989), "Exploration and Access to Public Lands", *Resources Policy*, Vol. 15, No. 2, June.

Geological Survey (1990), *Organization, Programs and Activities of the Geologic Division*, Washington D.C.

HOFFMAN, S. and HOUSEMAN, V. (1990), "Update on Regulations for Mine Waste Management under RCRA", *Mining Engineering*, November.

MARCUS, J. (1990), "Mining Environment — Regulatory Control of Mining at the Federal Level", *Engineering and Mining Journal*, June.

National Research Council (1990), *Competitiveness of the US Minerals and Metals Industry*, National Academy Press, Washington D.C.

Office of Technology Assessment (1985), *Strategic Materials: Technologies to Reduce US Import Vulnerability*, Washington D.C.

Office of Technology Assessment (1990), *Non Ferrous Metals: Industry Structure — Background Paper*, Washington D.C.

EUROPEAN COMMUNITY

INTRODUCTION

The European Community, as a whole, is the largest consumer of five of the six main non-ferrous metals and falls just behind the United States in aluminium consumption. On the production side, the EC holds significant shares for several non-ferrous metals. Thus, for both lead and zinc metal, it holds first place in the world. In the case of primary aluminium and refined copper production, the EC is collectively the second largest producer in the world.

As a rule, Community production tends to be concentrated in refined metal, with relatively lower shares of mine output. Nonetheless, at the mining stage, Community areas are important producers of bauxite (Greece), nickel (France/New Caledonia and Greece), lead (Spain, Ireland and Greece) and zinc (Spain and Ireland). Output of copper concentrates was expected to increase significantly as from 1990 (and tin in a second phase), when the operations of the Neves Corvo complex in Portugal reached full capacity. Special mention should also be made of the copper and copper alloy semi-manufactures in the Community, which represent nearly 40 per cent of world production of such semis.

It is estimated that mining employs about 25-30 000 people although this figure reaches over 300 000 when downstream processing and semi-manufacturing is included. Figures on employment have been included both in the individual Member States' Table 2 series and in Table 6.2.

In addition to being a large producer and consumer, the Community trades heavily in non-ferrous metals. Only in the case for zinc, however, is the Community a net exporter; for lead metal and products the Community exports and imports are approximately balanced. In the case of the other four metals, EC imports are large; in 1990, net trade in copper and in aluminium alone represented US$4.2 and 2.5 billion respectively in favour of non-EC States (cf. Table 6.3).

Finally, the importance of EC mining interests overseas is an essential background element in understanding developments in EC policies. As can be seen in Table 6.4, a large number of German, French, Dutch, Italian, Belgian and British companies have shares in bauxite, copper, lead, nickel, tin and zinc mining complexes abroad. The list would be longer were it to include other minerals, e.g. gold and uranium, or processing operations. The importance of EC-based companies overseas is such that, according to one analysis, the addition of the external interest (i.e. where processing plant feed comes from a mine with some EC participation) pushes the total EC interest over the 50 per cent mark for four of the six non-ferrous metals and to 40 per cent for alumina and tin[1].

The EC does not have a mining policy *stricto sensu*, in the way that it has, say, an agricultural policy or a fisheries policy. A certain number of activities carried out at the level of the Community and interventions by EC institutions do take place which impact on the sector, both within the Community as well as in developing countries.

In the 1980s the accession of Greece (1981) and Spain and Portugal (1986) caused a certain shift in perception of the centre of Community interests, to encompass not only those of "consumer" or "metal producer dependent on imported ores", but also that of "miner". This change in focus has in turn led to a movement of redefining policy, which is currently in an active phase. Indeed undeveloped resources, particularly in Ireland, Greece, Spain and Portugal, are deemed to represent potentially significant deposits of the main non-ferrous ores: e.g. Greece (bauxite, nickel, lead and zinc), Ireland (a few base metals plus gold), Portugal (copper and tin), and Spain (pyrites including copper and tin formations on the Portuguese border).

A decisive step was taken on 28 July 1989, when the EC Council adopted a resolution (see Annex at end of this chapter) setting out the parameters for a joint orientation in the use both of Member State and Community resources. Together with many of the activities under way in the context of completing the single market for 1993, it may be considered that the Community is moving towards, if not a formal policy, at least a common view as to the importance of the sector and ways to improve its long term competitiveness.

These are currently being reviewed at the technical level in four working groups composed of representatives of the Commission, Member States, and for some issues, with the non-ferrous mining and metals industry. While details of possible future programmes will undoubtedly take some time to be worked out, the main thrust of the problem areas identified in conjunction with the Council Resolution and the mandates of the working groups are sketched out below under relevant points. The following sections present a common view of main problems encountered and general lines of reflection for further joint examination between the Commission, Member States and the industry, rather than on officially adopted policy at this stage. In cases where existing policy or programmes are relevant, these have also been discussed.

PROSPECTING AND EXPLORATION

Geological services

The industry has emphasised the need for updating and ensuring better access to geophysical and general survey data for potential investors. From this perspective, more co-operation is called for between national geological surveys. One of the four working groups set up in early 1990 is dealing with the access and diffusion of geological and mining data. It is studying the operations of mining archives and the procedures for exchanging data on ore bodies within the EC and in certain third countries. In addition the whole matter of training and education needs in the fields of geology and mining engineering, including possible rationalisation and specialisation of mining schools is being looked at.

Exploration permits and concessions

Neither particular minerals nor specific areas should be reserved for state activities, in the view of the industry. The Commission underlined the fact that procedures for obtaining exploration licenses were heavy, slow and restrictive and may result also in uncertain legal rights. It is hoped through a joint examination of the problems, that the competent authorities, both national and sub-national, might adapt the legal environment to the current needs.

Taxation

The structure of tax incentives, such as in the form of depletion allowances, do not have a strong tradition in most European countries. Their absence, in the eyes of the industry has introduced distortions in prospecting effort away from European territory. Strong support is given for governments reviving tax incentives, such as exist in Canada, for attracting funds for the smaller or "junior" exploration companies, as well as a general tax deductibility of exploration costs. In its communication submitted to the Council, the Commission emphasised that the disparities in tax regimes for exploration expenses both among Community states and between EC and non-EC states justified a joint review by Member States and the Commission.

Direct participation

In December 1984 the EC Council reached general agreement concerning a regulation which had originally been proposed by the Commission in 1982, setting up a scheme for exploration loans[2]. Under such a scheme, between 30 and 50 per cent of planned spending would be on-lent from Community funds, complementing national funds where these were insufficient, but still leaving the bulk to be borne by private operators; subsidies would be repaid in the case of exploitation of a successful find. The proposed support of prospecting and drilling would extend to Community territory only. However, attempts in the 1980s to fund the scheme never succeeded, due to low metal prices, budgetary difficulties and lack of interest on the part of certain Member States. The Commission argues that the situation has now evolved, with the accession of Greece, Spain and Portugal, and notably due to their mining potential, it recommends that part of the Structural Funds' financial resources be affected to mining projects.

MINING AND METALS PRODUCTION

Regulations

EC industry emphasises the cumulative nature of various sets of national and regional regulations building up over the years concerning social, environmental and sovereignty over natural resources. The result in their opinion has been complex and burdensome requirements and consequent delays. It is felt that a unanimous and consistent willingness is needed to facilitate the creation of new mines. A special case in point concerns social charges. Charges in this sector were inherited from a period when mining was a much more important activity than it is today and are often no longer justified. The Commission has expressed the concern that authorisation for opening mines may be spread out over several years. Administrative procedures, they agree, often do not only not favour the opening of new mines, but impede the closing of mines which are no longer viable.

Environment

Industry complains of long delays in obtaining access to mining rights due to environmental regulations and feels that the mining industry is today, by definition almost, totally sensitised and aware of the fragility of its working ecology. Emphasis should be on common interpretation and implementation of EC laws and regulations; further, new provisions should be based on scientific and technological findings.

EC environmental policy is still relatively young. While nothing appears explicitly in this regard in the Rome Treaty, a series of action programmes have been adopted in the 1970s and 1980s. In the mid-1980s emphasis was on specific measures rather than elaborating a comprehensive and coherent environment policy. Nonetheless, in December 1985, the European Council reached agreement in principle on the extension of Community competence to the field of the environment, in cases where protection is better reached at the Community level than at national level, and with decisions taken unanimously. The Single European Act, which entered into force in 1987, contains three articles specifically on the environment, as well as a reference in relation to competition[3]. Since then, a number of Directives have been adopted.

One of these, 85/337/EEC, deals with environmental impact assessments (EIA) on certain public and private projects[4]. States must ensure assessment is undertaken on projects on a specified list. While non-ferrous mining and metals works do not appear on the obligatory list (except for asbestos), Annex II concerns projects which are subject to an EIA "where Member States consider their characteristics so require". The mining industry is one of those so listed. Procedures include evaluating direct and indirect effects on water, fauna and flora, air, surrounding landscape, etc. Information must be available to the public and an opportunity given to the public concerned to express an opinion before a decision is taken. This Directive is currently being followed up at the national and/or regional level by implementing regulations.

The Commission is of the view that the conditions for simplification and better co-ordination of the authorisation procedures vis-à-vis mining installations exist in this decree. In the current phase of discussions, a separate working group has been set up to review all regulations dealing with exploration and mining, including EC and national standards on the protection of the environment.

Minerals taxation

In addition to the particular taxation problems related to exploration mentioned above, industry and the Commission agree that national tax legislation is not well suited to the life of mines in respect to amortisation and the carrying-over of losses during the first years of activity. The nature of mining being one of exhausting its own capital, the possibility of a depletion allowance — as some important mining countries incorporate in their tax codes — should be more actively encouraged. Special tax provisions, such as allowable deductions are generally not generous or non-existent, including for costs associated with site clean-up and reclamation costs at the end of the life of a mine. One of the joint working groups is currently dealing with the various taxation issues.

Inputs

Within the context of implementing the Single Act, the Commission is working towards liberalisation and efficient use of investment in the energy sector, involving the improvement of price transparency and transit arrangements for gas and electricity. It is expected that in time this could have significant impact on national power monopolies.

Direct and indirect support

As mentioned above, the Commission is not supporting the call of certain Member States to create a special mining fund, feeling that existing Community Structural Funds were sufficient when financing was needed to support interventions in non-ferrous mining and metal activities. This would necessitate interested Member States earmarking a small portion of such funds for sectoral projects.

The Structural Funds in question are the European Regional Development Fund (ERDF), the European Investment Bank (EIB) and the European Social Fund (ESF). ERDF contributed to 39 mining investments between 1985 and 1988 for a value of ECU 10.6 million in the less-favoured regions of Denmark, Western Germany (Federal Republic of Germany before the unification of Germany), Italy, Greece, Spain and the UK. Examples of EIB involvement include the financing of the tin and copper mine of Neves Corvo in Portugal. The European Social Fund has been used in the context of mine closures.

In discussing a new policy, the Commission has stressed that Member States that so wish have at their disposal potential financial support in the Structural Funds which may be used to further development of their mining sector. Within the goal imposed by the Single European Act of increasing economic and social cohesiveness among Community regions, the recent reform of the Structural Funds places great emphasis on co-ordination and multi year programming between the Commission, the various Funds and the EIB and within established Community support frameworks in partnership with relevant Member State authorities.

Some national programmes are of interest to the Community and thus eligible to certain measures of financial support by the ERDF, either directly as Community initiatives or from funds made directly available to Member States. Current projects under consideration for support include the restructuring of a bauxite mine and a alumina plant in Greece, and the evaluation of pyrites mines in the Spanish-Portuguese transfrontier region.

HORIZONTAL POLICIES

Competition policy

The Commission has considerable powers in the field of competition policy. In particular, under Articles 92-94 of the Treaty of Rome, the Commission has the power to grant a derogation to an otherwise general ban on state aids. It therefore has a wide margin of appreciation to determine if the aid in question promotes a recognised Community goal without distorting competition to an extent contrary to the common interest. Therefore before any aid scheme is put into operation by a Member State it must be notified to the Commission. The Commission also has the power to review the operation of any existing aid scheme and demand changes if it considers them necessary. It may even demand that a company repay an aid if it has been given without its approval.

Over time, the Commission has developed a policy for the different types of aids. For certain sectoral aids, an almost blanket ban is in force, e.g. on synthetic fibres. On the basis of certain Community objectives, such as regional development of less favoured regions or promotion of small and medium enterprises, aids are looked at in a more tolerant way, certain schemes being permitted up to a predetermined level of intensity. In the area of non-ferrous mining and metals, the Commission has taken a case-by-case approach, allowing certain state injections of money, while disallowing others. A few recent examples are mentioned below, purely in an illustrative fashion.

In 1988, the Commission cleared aid given by the French Government to Péchiney in the amount of over ECU 1 billion through new capital, participating loans and a loan from disposing a subsidiary. These infusions took place in the context of a major restructuring programme which included disposal of certain interests, financing of redundancy costs, and the creation of alternative jobs.

Approval had been given on the stipulation that no aid to increase capacity was given before the end of 1990.

In May 1989, a negative decision was rendered concerning Italian state aids (ECU 66 million) going to Aluminia and Comsal, Italian state aluminium producers. The negative decision was reached as the aid had not been previously notified and violated a condition attached to Commission approval of previous support to restructuring of the public aluminium industry over the 1983-88 period.

In June 1989, a procedure was opened in respect of support foreseen in the new Italian five-year plan on mining policy. Although the Commission did not raise objections concerning the granting of subsidies on basic research and applied research, nor for exploration abroad, continued support to restructuring activities and the size of environmental aids were considered as possible distortions to competition. Much of this aid was directed at non-ferrous activities in Sardinia. The procedure was closed in February 1990, after the amount of subsidies were, it is reported, pruned back by Italian authorities.

In 1989, the Commission again investigated planned French support to Péchiney for the new smelter at Dunkirk. The controversial point concerned the terms of the power contract, whereby, according to competing German and British firms, electricity was to be furnished below its real cost. Approval was finally granted after it was agreed to increase the cost of electricity (see chapter on France for details).

In 1989, the Commission approved continuation of Portuguese aid to certain tin and tungsten mines. Subsidies cover maintenance costs and unemployment benefits paid for the technical work stoppage. They are however, to be repaid over a five year period, in the case that the mines start producing again.

In December 1989, the Commission closed its procedure concerning a Greek subsidy of ECU 15.6 million to the bauxite mining company Eleusis. The form of aid, both of converting debts into capital and injecting new capital from the government did not respect obligations laid down by the Commission in an earlier decision on a Greek financing agency set up to help enterprises in difficulty. In taking a favourable attitude on the aid, the Commission took into account the EC's deficit in bauxite and the low degree of competition in the Community, as well as additional restructuring measures decided on for the enterprise in question.

Research and development (R&D)

As set forth in the European Single Act, governments research and technological development operates within so-called framework programmes, i.e. a programming instrument which defines, in a multi-year perspective, the entire set of R&D activities of the Community. Political agreement was reached recently in the Council on the Commission's proposal for a third framework programme for the five-year period 1990-94 with proposed funding of ECU 5.7 billion. EC Framework programmes are "rolling", a proposal for a future programme overlaps with the current one in order to avoid sudden shifts in direction. Thus the former programme (1987-91) is still being implemented. In June 1990 the Council approved a new R&D programme in the field of raw materials and recycling (1990-92), with a proposed budget of ECU 45 million and it is this programme which is briefly discussed here.

The general aims of the programme are to enhance the competitive position of the EC industries involved with raw materials and recycling and to help provide the technology base required to improve the supply and processing of raw materials and recycling. The two main parts of the programme are sub-divided into three technical areas as follows: exploration; mining technology; and mineral processing and extractive metallurgy.

Research contracts are mainly on a shared cost basis, i.e. the Commission finances up to 50 per cent of total costs. The partner may either be from the public sector, such as a university or research institute, or from industry, particularly small and medium-sized enterprises. Projects usually involve a consortium of companies and institutions.

Trade policy

For the three non-ferrous metals in which the Community has the greatest production interests — aluminium, lead and zinc — the EC's external tariff shows a classical pattern of escalation. Ores and concentrates enjoy free entry; unwrought metals face low level duties (3.5-6 per cent) and duties on wrought products range from 7.5 to 10 per cent (see Table 5.2). Weighted averages as calculated by the GATT (see Table 4.2) were 9.8, 7.8 and 7.7 per cent respectively for semis of aluminium, lead and zinc. In the cases of nickel and tin, where Community mining, smelting and refining interests are much smaller, not only ores and concentrates, but unwrought metal faces zero or very low duties, and tariffs on wrought products are at low levels (2.5-5.3 per cent). Tariffs on unwrought copper were zero before the Tokyo Round, as the bulk of copper metal imports are industry inputs, destined for further processing (mainly in Western Germany and Belgium). Tariff duties on copper semis range between 5 and 6.5 per cent.

Quantitative restrictions which were applied by some Member States on imports of unwrought metals (aluminium and zinc), and to some semis (aluminium, lead and zinc) are being progressively eliminated.

Recent trade liberalisation discussions in the non-ferrous area have focussed on the proposed preferential duties for aluminium metal from the Gulf States in the context of a free trade agreement. In view of important capacity being installed or under construction there, elimination of the 6 per cent tariff on aluminium ingot increases competition for ingot manufacturers located within the EC. The Community has recently granted GSP status to Poland and Hungary (01.01.90) and to Bulgaria and Czechoslovakia (01.01.91), whilst Romania had benefited therefrom since 1974. Regarding non-ferrous products, this has meant larger preferential margins for exporters in the eastern European countries of copper and aluminium semis.

A series of anti-dumping investigations have been carried out in recent years, most of them concerning ferro-alloys or other steel inputs, including tungsten, silicon, boron, titanium, and antimony, but — in recent years — none on the six metals under review. The fist two led to imposition of an anti-dumping duty on imports from China.

Export controls on non-ferrous scrap had been in place through the 1970s and 1980s. While these only consisted of general surveillance for aluminium and lead scrap, export quotas were established each year by the Council of Ministers for copper and copper alloy scrap. In December 1989, it was decided not to renew the system for 1990, so that exports of copper scrap are totally free.

CONSUMPTION/SUPPLY

Securing overseas supply

In the late seventies with preoccupations in the Community focussing on adequacy of investment and sufficiency of supply of non-fuel minerals and at a time when the North-South dialogue was a particularly important element in policy formulation, the possibilities for extending STABEX to cover minerals were explored. Under the first Lomé Convention, iron ore had been included, but the sheer size and importance of trade in non-fuel minerals did not however permit such an extension of coverage.

Instead, under Lomé II (1980-85) a special facility for minerals was introduced, referred to as SYSMIN. Its funds, amounting to ECU 282 million, (and increased to ECU 415 million under Lomé III) were available to African, Caribbean and Pacific (ACP) countries for which the exports of certain non-fuel minerals (copper and cobalt, phosphates, manganese, tin, bauxite and alumina, or iron ore) represented at least 15 per cent of total exports or where all mineral production reached 20 per cent of total exports (fuels and precious minerals excluded). While this new facility was primarily conceived as an aid mechanism in view of the fragility of the mineral economies of the ACP, it was also clearly designed as an instrument to assist in securing supplies of raw materials for European industry. In the early years of SYSMIN, interventions were specifically limited to the extractive and beneficiation stages, although not generally for increasing output (cf. Table 6.5). Early Conventions were geared to safeguarding production and export capacity. While interventions were triggered by a fall in export receipts (of at least 10 per cent in comparison with the previous four years) the system was never designed as a system of compensatory financing. The accent was placed on programmes of rehabilitation or rationalisation with a view to keeping up the viability of the sector.

Financing has not taken the form of simple budget compensation but is project aid and often involves co-financing with other institutions. Thus considerable delays were encountered under the first two five-year exercises and the total amount programmed has not been disbursed. As can be seen from

Table 6.5, under Lomé III only one-fourth was committed and only one-tenth of this had been disbursed.

In the light of these considerations, several innovations have been introduced into the revised version of SYSMIN of Lomé IV. For the first five-year period, the financial package has increased by some 15 per cent to ECU 480 million. While it essentially remains project aid in nature, the scope of possible activity has been widened. Within the general objective of safeguarding production and export capacity, financial restructuring may now be undertaken. The field of possible interventions aiming at diversification, is considerably widened since financing can go to economic activities completely outside the mining sector when a return to viability is not considered possible or the economy's dependence on minerals is too high. In addition, the Convention henceforth permits "preventive" actions, i.e. financing of maintenance projects or others which would help to ensure viability in the sector.

The number of countries potentially affected has been increased since uranium and gold have been added to the list of products for which a drop in earnings may trigger eligibility. The dependency thresholds remain the same, at 15 per cent for any mineral on the list of eligible products or a collective 20 per cent dependency on overall exports of products on the list. These shares are reduced to 10 and 12 per cent respectively for the ACP least-developed, island and land-locked countries.

It should also be noted that Lomé IV has strengthened considerably the requirement that SYSMIN interventions in ACP producing countries not damage Community producers of the same commodity. This has been a sore point in the past, particularly for certain bauxite producers, which considered that their own mines were threatened by Community aid to ACP mineral countries.

Recycling

In the general area of recycling, the Community has two sub-programmes of research and development, the first entitled recycling of non-ferrous metals and strategic metals with the following four technical areas:

a) characterisation and classification of secondary materials and physical separation and concentration;

b) advanced pyrometallurgical processes;

c) advanced hydrometallurgical processes;

d) refining technologies and instrumentation on control of the processes.

In addition, a further sub-programme called REWARD, deals with recycling of domestic waste. The Community also has an important research activity on advanced materials called EURAM — European Research on Advanced Materials — with funding over the 1987-91 programme of ECU 220 million.

Resolution on the development of the Community mining industry (28 July 1989)

The Council of the European Communities,

Having regard to the EEC Treaty,

Recognising that the mining industry is of significant economic, social and commercial importance and can make an appreciable contribution to the provision of Community supplies by its activities both within the Community and in third countries;

Considering that the mining sector must, like other sectors, be integrated into the plan for the completion of the internal market in 1993 by strengthening its production structures through improvements in competitiveness;

Noting that development of the mining sector is bound to have a positive effect on an entire industrial network including a considerable mining equipment and services sector upstream and all of the processing industries downstream;

Believing that the Community mining industry must be developed under normal conditions of competition and with a firm Community commitment to liberalisation of access to resources world-wide;

Welcomes the memorandum on the mining industry submitted by the Commission and recognises the need for a common approach by the Community and in particular for co-ordination of national and Community measures and budgetary instruments assigned to industrial development;

Calls on the Commission, in consultation with the Member States, to continue without delay with preparation of the specific measures contained in the action envisaged in the memorandum, having regard to the relevant Community provisions;

Emphasises that in carrying out specific measures priority should be given to the following points:

1. on R&D, any adjustment of current programmes such as the framework programmes must follow from an examination of national efforts, rigorous analysis of actual requirements and identification of the most appropriate Community action;
2. the Structural Funds' existing allocations and the EIB's loans could, if necessary, provide as far as possible the necessary assistance for the

development of mining projects of Community scope, including exploration projects;

3. with the completion of the internal market in prospect, a study should be made by the Member States and the Commission with the aim of eliminating administrative and/or fiscal obstacles and improving the socio-economic framework for the sector;

4. particular attention should be paid to examining training and education requirements in the field of geology and mining engineering, with combined action possibly being sought between Member States and also with major mining third countries.

NOTES AND REFERENCES

1. D. HUMPHREYS, "Mining Interests Overseas", *Resources Policy*, March 1988.
2. *EC Bulletin*, 12-1984 and *Official Journal* C203, 6 August 1982.
3. Articles 130 (R-T) and Article 100, respectively.
4. 85/337/EEC in *Official Journal*, L 175, 5 July 1985.

BIBLIOGRAPHY

Commission of the European Communities, COM(90)278 final, 2 June 1989.

Commission of the European Communities, *Report on Competition Policy*, various issues.

Commission of the European Communities, *General Activities Report*, various years.

Eurométaux(1989), *Conditions of Profitability of the EC Mining Industry*, November .

Eurométaux (1990), *Raw Materials for Europe*, January.

HUMPHREYS, D. (1990), "Towards an EEC minerals policy?", *Resources Policy*, March.

ANNEX

TABLES

Table 1.1 Incentives for exploration activities and direct government participation

Country	Tax deduction percentage	Other incentives and direct government participation
Australia	100%	Each state has its own incentives.
Canada	100%	Each province has its own incentives. Up to 1990, exploration expenses may be deducted by investors under a programme of "flow-through shares".
France		Tax relief for constituting "provisions" for future investments in exploration. BRGM carries out exploration on national territory and abroad.
Germany	100%	Exploration costs accounted as operating costs.
Greece	100%	Responsibility for conducting basic geological exploration and mineral deposit surveys is vested in the state, through IGME.
Ireland	100%	Prescribed exploration costs treated as capital expenditures. 20% tax allowance abolished in 1990. Grants for up to 50% of prescribed costs, repayable in case of finds.
Italy		Grants up to 70% of costs of exploration on national territory or abroad; state loans covering the remaining 30%, reimbursable in case of commercial finds.
Japan		Government finances 50% of exploration costs of new deposits incurred by small firms and provides subsidised loans on up to 80% of costs incurred by large firms for exploration projects in remote areas (for Cu, Pb, Zn, Mn, Au, Mo, W).
Norway	100%	Exploration expenditures are normally fully deductible from taxable income in the year of accrual. Government finances up to 50 per cent of exploration costs of new deposits. The total government expenditure in that respect has so far been limited to NKr4 million per year. For tax purposes investments are capitalised and depreciated according to the ordinary rules in the tax system.
Portugal	100%	Exploration costs may be treated as fixed assets and have applied depreciation rate of up to 33.3%.
Spain		Grants for up to 20% of exploration costs, repayable in case of finds.
Sweden	100% (through 1991)	Through its own agency, the state conducts on average one third of exploration investment.
United Kingdom	25%	When exploration leads to trade in mineral, then expenditures may be treated as capital expenditures.
United States	70%	Remaining 30% is capitalised and depreciated over 5 years.

Table 1.2 **Taxation applied to mining activities**

Country	Basic (corporation) rate	Royalties	Other taxes	Exemptions
Australia	39%	Royalties by mineral. Vary according to state	Other charges	Tax exemption on gold production until 1991. A few infrastructure investments written off in 10 years
Belgium	39%			
Canada	28%	Royalties vary by province	Local taxes	Accelerated depreciation for certain assets, environment and equipment leading to lower energy consumption. Provincial income and mining taxes produce effective overall tax rates ranging from 33 to 51 per cent
France	34% (base rate; reinvested profits) +8% (distributed profits)			Accelerated depreciation provision for "reconstitution" of the ore deposit (i.e. a depletion allowance)
Germany	36% (distributed profits) 50% (undistri-buted profits)			
Greece	40% (distributed profits) 35% (reinvested profits)			
Ireland	40%	Royalties or percentage profit to private bodies or the state		
Italy	25% +11.25%	Royalties fixed case by case		
Japan	37.5% (+12.0)		Local taxes	Mining firms are exempt from the 12.0% enterprise tax. There is accelerated depreciation for certain investments. Depletion allowances in force up to 1991 leading to the deduction of up to 50% of gross income from taxable income or 13% of sales
Norway	28%			There are no specific rules for the mining sector
Portugal	36.5%	Royalties	4% (sales)	
Spain	35%			Exemptions up to 100% from several taxes for a limited period applying to priority activities

Table 1.2 (cont'd) **Taxation applied to mining activities**

Country	Basic (corporation) rate	Royalties	Other taxes	Exemptions
Sweden	30%			Several exemptions for underground mines
United Kingdom	33% (base) 25% (SME for profits under £25 000)	Royalties to land owner		
United States	34%		0 to 12% depending on the state; average 6.5%	Possible reduction (up to 50% of net profits) of part of the investment or sales costs; percentage depletion allowances: 15% (copper, silver) or 22% (nickel, lead, zinc)

Table 1.3 **MFN tariffs on three non-ferrous ores and metals (*ad valorem* rates, unless otherwise specified)**

Country Non-ferrous metal	USA	EEC	Japan	Canada	Sweden	Australia	Norway
Ores							
Zinc	(1)	0	0	0	0	0	0
Copper	(2)	0	0	0	0	0	0
Aluminium	0	0	0	0	0	0	0
Unwrought							
Zinc(3)	1.5-19.0	3.5	(4)	0-17.5	0	0	0
Copper(5)	1.0	0	(6)	0-10.3	0	0	0
Alum.(7)	0-2.6	6.0	1.0	0	0	0	0
Wrought(8)							
Zinc	4.2	8.0	3.8	0-10.2	0	0	0-3.8
Copper	1.0-6.3	6.0	4.6-5.8	4.0-10.3	1.4-2.5	10.0	0-5.1
Aluminium	2.6-5.0	10.0	9.2	2.1-10.3	2.5	10.0	0-13.0

1) 0.7¢/kg on copper content plus 1.7¢/kg on lead content.
2) 1.7¢/Kg on lead content.
3) Unwrought zinc HS 79.01.
4) Japan applies a sliding tariff system in which the specific rate decreases with the increase in the price of refined zinc. The bound rate is 8 yen/kg.
5) Refined copper and copper alloys, unwrought HS 74.03.
6) Japan applies a sliding tariff system in which the specific rate decreases with the increase in the price of refined copper. The bound rate is 21 yen/kg. Detailed explanation of the sliding tariff is given in the chapter on Japan.
7) Unwrought aluminium HS 76.01 (76.01.10.10 and 76.01.20.10 for Canada).
8) Zinc bars, rods, profiles and wire HS 79.04; copper bars, rods and profiles HS 74.07; aluminium bars, rods and profiles HS 76.04.10; rates indicated for Japan are applied rates which are lower than MFN rates.

Source: GATT and country chapters.

Table 2.1 **Australia: production/consumption, trade and employment in the non-ferrous metals sector, average 1990/1991 ('000 metric tonnes)**

	Production	Imports	Exports	Consumption
Bauxite	40 947.0 [a]			
Alumina	11 472.0			
Aluminium	1 258.3	2.0 [b]	930.7[b]	300.5
of which primary	1 230.6			
of which secondary	27.7			
Aluminium semi-manuf. [b]	275.3	19.3	57.2	
Copper ores/concs.	323.5			
Copper blister	203.6 [c]		4.0[c]	
Copper refined	276.5	1.1	162.1	110.6
Copper semi-manuf.	60.2 [d]	14.5	26.5	
Nickel ores/concs.	68.0			
Nickel refined	46.3			1.5
Lead ores/concs.	574.5		83.1[e]	
Lead refined	234.2		176.1	57.0
of which secondary	18.2			
Zinc ores/concs.	994.0 [f]		659.0[f]	
Zinc slab	317.5		175.2	155.3
Tin ores/concs.	6.6			
Tin refined	0.6		0.3	1.3
Employment in non-ferrous mining and metals ('000)		62 (1989)		
as a percentage of total labour force [g]		0.7%		
Exports of non-ferrous ores/metals (million US$)		3 806.2		
as a percentage of total exports [h]		11.1%		
Imports of non-ferrous ores/metals (million US$)		246.8		
as a percentage of total imports [h]		0.6%		

[a] Gross weight
[b] Including aluminium alloys
[c] Including copper anodes
[d] Including copper alloy semis
[e] Lead content
[f] Zinc content
[g] Total employment in metal ore mining and non-ferrous metals manufacturing (ISIC 230 and 372)
[h] 1989/1990

Source: World Metal Statistics, World Bureau of Metal Statistics; Metallstatistik 1980-1990, Metallgesellschaft AG; Statistical Yearbook 1990, Metaleurop; Metalli Non Ferrosi 1991, Nuova SAMIM; Foreign trade data, OECD; Industrial Statistics Yearbook 1989, UN; Yearbook of Labour Statistics 1991, ILO.

Table 2.2 **Belgium: production/consumption, trade and employment in the non-ferrous metals sector, average 1990/1991 ('000 metric tonnes)**

	Production	Imports	Exports	Consumption
Bauxite		31.2 [a,b]	1.7 [a,b]	
Alumina		10.9 [a,b]	1.4 [a,b]	
Aluminium [c]		369.1	43.3	
of which primary				320.4
of which secondary	3.0			
Aluminium semi-manuf. [d]	340.4	141.9	359.9	
Copper ores/concs.		4.1 [e]		
Copper blister	87.7	176.7 [f]		
Copper refined	314.7 [g]	248.4 [h]	206.3 [a,f]	380.8
Copper semi-manuf. [d]	379.5	58.3	332.2	
Nickel refined				20.6
Lead ores/concs.		46.0 [a,b]		
Lead refined	95.7	21.2	46.2	67.6
of which secondary	21.1			
Zinc slab	293.7	37.2	133.8	188.8
Tin refined	0.2	2.4	2.8	1.5
Employment in metals manufacturing ('000)		64 (1985)		
as a percentage of total labour force [i]		1.5%		
Exports of non-ferrous ores/metals (million US$)		3 360.3 [a]		
as a percentage of total exports		2.8%		
Imports of non-ferrous ores/metals (million US$)		3 331.6 [a]		
as a percentage of total imports		2.8%		

[a] Including Luxembourg
[b] 1989/1990
[c] Including aluminium alloys
[d] Including alloy semis
[e] Gross weight
[f] Including copper anodes
[g] Estimates
[h] Including leach cathodes
[i] Total employment in metals manufacturing [ISIC 371 (fe/steel) & 372 (non-ferrous)]

Source: World Metal Statistics, World Bureau of Metal Statistics; Metallstatistik 1980-1990, Metallgesellschaft AG; Statistical Yearbook 1990, Metaleurop; Metalli Non Ferrossi 1991, Nuova SAMIM; Foreign trade data, OECD; Industrial Statistics Yearbook 1989, UN; Yearbook of Labour Statistics 1991, ILO.

Table 2.3 **Canada: production/consumption, trade and employment in the non-ferrous metals sector, average 1990/1991 ('000 metric tonnes)**

	Production	Imports	Exports	Consumption
Bauxite		2 425.6 [a]	2.3 [a]	
Alumina	1 109.7	2 002.7 [a]		
Aluminium	1 762.2	30.0 [b]	1 361.4 [b]	465.3
of which primary	1 694.5			397.7
of which secondary	67.7			67.7
Copper ores/concs. [c]	795.6	51.0	334.2	
Copper blister [d]	534.7			
Copper refined	527.1	2.4	353.7 [e]	184.8
Copper semi-manuf.	130.9	61.1 [f]	23.8 [f]	
Nickel ores/concs. [g]	196.6	0.0	40.1	
Nickel refined	123.5 [h]	6.5	72.5	14.5 [h]
Lead ores/concs. [i]	259.7		621.3	
Lead refined	197.5	9.0	90.0	77.7
of which secondary	100.7			
Zinc ores/concs.	1 175.7 [j]	174.0	727.2 [j]	
Zinc slab	626.2	2.4	484.9	122.0
Tin ores/concs.	4.2			
Tin refined	0.1	3.4		3.4

Employment in non-ferrous mining and metals ('000)	95 (1988)
as a percentage of total labour force [k]	0.7%
Exports of non-ferrous ores/metals (million US$)	7 666.1 [a]
as a percentage of total exports	6.6%
Imports of non-ferrous ores/metals (million US$)	2 446.4 [a]
as a percentage of total imports	2.1%

[a] 1989/1990
[b] Including aluminium alloys
[c] Copper content
[d] Including copper anodes
[e] Including some blister
[f] Including alloy semis
[g] Nickel content
[h] Smelter refinery products (Nickel content)
[i] Lead content
[j] Zinc content
[k] Total employment in metal ore mining and non-ferrous metals manufacturing (ISIC 230 and 372)

Source: World Metal Statistics; World Bureau of Metal Statistics; Metallstatistik 1980-1990, Metallgesellschaft AG; Statistical Yearbook 1990, Metaleurop; Metalli Non Ferrosi 1991, Nuova SAMIM; Foreign trade data, OECD; Industrial Statistics Yearbook 1989, UN; Yearbook of Labour Statistics 1991, ILO.

Table 2.4 **France: production/consumption, trade and employment in the non-ferrous metals sector, average 1990/1991 ('000 metric tonnes)**

	Production	Imports	Exports	Consumption
Bauxite	336.6 [a]	1 368.7 [b]	21.0 [b]	
Alumina	572.0	412.3 [b,c]	235.6 [b,c]	
Aluminium	519.0	546.3 [d]	131.7 [d]	981.2 [b]
of which primary	306.0			727.6
of which secondary	213.0			
Aluminium semi-manuf.	728.3 [e]	134.0	75.9	
Copper ores/concs.	0.3		1.1 [b]	
Copper blister [f]	6.2	16.8	4.5	
Copper refined	53.8	433.2	9.8	479.4
Copper semi-manuf. [e]	744.6	263.0	349.9	
Nickel ores/concs. [g]	99.8		42.4	
Nickel refined	9.2	24.6	5.5	40.8
Lead ores/concs.	1.5	179.0 [a]		
Lead refined	271.6	42.0	53.7	253.5
of which secondary	146.2			
Zinc ores/concs.	25.5	613.8 [a]	18.4	
Zinc slab	281.9	85.6	77.6	286.6
Tin refined		8.3	0.1 [b]	8.3
Employment in non-ferrous mining and metals ('000)		56 (1989)		
as percentage of total labour force		0.3%		
Exports of non-ferrous ores/metals (million US$)		3 366.6		
as percentage of total exports		1.6%		
Imports of non-ferrous ores/metals (million US$)		5 804.2		
as percentage of total imports		2.5%		

[a] Gross weight
[b] 1989/1990
[c] Including alumina hydrates
[d] Including aluminium alloys
[e] Including alloy semis
[f] Including copper anodes
[g] New Caledonia

Source: World Metal Statistics, World Bureau of Metal Statistics; Metallstatistik 1980-1990, Metallgesellschaft AG; Statistical Yearbook 1990, Metaleurop; Metalli Non Ferrossi 1991, Nuova SAMIM; Foreign trade data, OECD; Industrial Statistics Yearbook 1989, UN; Yearbook of Labour Statistics 1991, ILO.

Table 2.5.1 **Germany: Western Germany — production/consumption, employment and trade in the non-ferrous metals sector, average 1990/1991 ('000 metric tonnes)**

	Production	Imports	Exports	Consumption
Bauxite		2 978.1 [a]	73.0 [a]	
Alumina	1 160.0	1 001.6 [a]		
Aluminium	1 244.0	952.8 [b]	302.0 [b]	1 891.9
of which primary	705.3			1 328.2
of which secondary	538.7			563.8
Aluminium semi-manuf. [c]	1 373.2	579.8	620.3	
Copper ores/concs.	2.0	505.4 [d]	0.1 [d]	
Copper blister [e]	253.4	56.4	0.5	
Copper refined	498.9	535.5	76.2	945.8
Copper semi-manuf. [c]	1 373.0	352.3	531.7	
Nickel refined		56.0 [a]	4.7 [a]	82.9
Lead ores/concs.	8.0	190.9 [d]		
Lead refined	355.6	107.1	61.6	402.7
of which secondary	194.2			
Zinc ores/concs.	56.0	608.7 [d]	33.5 [d]	
Zinc slab	341.7	246.3	74.2	512.1
Tin ores/concs.		0.4 [a,d]		
Tin refined	0.6	19.6	1.0	19.0
Employment in non-ferrous mining and metals ('000)		104 (1989)		
as a percentage of total labour force [f]		0.4%		
Exports of non-ferrous ores/metals (million US$)		6 576.9		
as a percentage of total exports		1.6%		
Imports of non-ferrous ores/metals (million US$)		9 270.5		
as a percentage of total imports		2.5%		

[a] 1989/1990
[b] Including alloys
[c] Including alloy semis
[d] Gross weight
[e] Including copper anodes
[f] Total employment in metal ore mining and non-ferrous metals manufacturing (ISIC 230 and 372)

Source: World Metal Statistics, World Bureau of Metal Statistics; Metallstatistik 1980-1990, Metallgesellschaft AG; Statistical Yearbook 1990, Metaleurop; Metalli Non Ferrossi 1991, Nuova SAMIM; Foreign trade data, OECD; Industrial Statistics Yearbook 1989, UN; Yearbook of Labour Statistics 1991, ILO.

Table 2.5.2 **Germany: New Federal Länder — production/consumption and trade in the non-ferrous metals sector, average 1989/1990 ('000 metric tonnes)**

	Production	Imports	Exports	Consumption
Bauxite		152.9		
Aluminium and alloys [a]	53.9	184.8	5.5	222.4
Aluminium semi-manufactures and alloys [b]	184.6	11.5	24.2	170.6
Cast products from aluminium and alloys	66.5	2.4	0.5	68.3
Copper ore/concs. [c]	7.4			
Refined copper and electrolyte copper	44.4	89.2		131.9
Copper semi-manufactures	101 (*)	0.9	18.9	83.4
Nickel & ferro-alloys, electrolyte nickel & nickel sulphate [d]	1.9			
Lead refined	40.9	39.2	0.5	76.4
Tin ores/concs. [e]	2.4			

[a] Including primary materials
[b] Excluding primary materials for semi-manufactures
[c] 1988/89: silver 53t metal content
[d] Estimated metal content
[e] Tin in pure ore concentrate
(*) Recalculated (printing error in the official statistics)
Source: Yearbook of Mining, Oil and Gas, Electricity and Chemicals, 1991; Statistical Yearbook of GDR Industry 1990, Federal Statistics Office; Statistical Yearbook 1990 for the FRG, Wiesbaden, September 1990.

Table 2.6 **Greece: production/consumption, trade and employment in the non-ferrous metals sector, average 1990/1991 ('000 metric tonnes)**

	Production	Imports	Exports	Consumption
Bauxite	2 314.7 [a]			
Alumina	613.0 [b]			
Aluminium	151.1	19.2	61.4	116.7
Aluminium semi-manuf.	123.0			
Copper refined		50.0 [c]		50.4
Copper semi-manuf.	70.7 [d]			
Nickel ores/concs.	18.3			
Nickel refined	15.9			0.2
Lead ores/concs.	29.0			
Lead refined		13.2	0.1	15.9
Zinc ores/concs.	28.4			
Zinc slab		11.7 [e]		17.9
Tin refined	0.2			0.9
Employment in non-ferrous mining and metals ('000)		9.7 (1986)		
as a percentage of total labour force [f]		0.3%		
Exports of non-ferrous ores/metals (million US$)		460.7		
as a percentage of total exports		5.5%		
Imports of non-ferrous ores/metals (million US$)		353.9		
as a percentage of total imports [g]		1.8%		

[a] Gross weight
[b] Alumina hydrate
[c] Primary products only
[d] Including alloy semis
[e] Including alloys
[f] Total employment in metal ore mining and non-ferrous metals manufacturing (ISIC 230 and 372)
[g] 1990

Source: World Metal Statistics, World Bureau of Metal Statistics; Metallstatistik 1980-1990, Metallgesellschaft AG; Statistical Yearbook 1990, Metaleurop; Metalli Non Ferrosi 1991, Nuova SAMIM; Foreign trade data, OECD; Industrial Statistics Yearbook 1989, UN; Yearbook of Labour Statistics 1991, ILO.

Table 2.7 **Ireland: production/consumption, trade and employment in the non-ferrous metals sector, average 1990/1991 ('000 metric tonnes)**

	Production	Imports	Exports	Consumption
Bauxite		1 606.2 [a]		
Alumina	953.5 [b]			
Aluminium		7.7	1.8	
Copper refined				0.2
Nickel refined				0.1
Lead ores/concs.	37.6			
Lead refined	13.3			20.6
of which secondary	13.3			
Zinc ores/concs.	167.7		173.5	
Zinc slab		1.3 [c,d]	0.1	1.4
Tin refined				0.1

Employment in non-ferrous mining and metals ('000)	1.3 (1986)
as a percentage of total labour force [e]	0.1%
Exports of non-ferrous ores/metals (million US$)	402.7
as a percentage of total exports	1.7%
Imports of non-ferrous ores/metals (million US$)	374.7
as a percentage of total imports	1.8%

[a] Gross weight
[b] Alumina hydrate
[c] Including alloys
[d] 1990, Jan. to Nov.
[e] Total employment in metal ore mining and non-ferrous metals manufacturing (ISIC 230 and 372)

Source: World Metal Statistics, World Bureau of Metal Statistics; Metallstatistik 1980-1990, Metallgesellschaft AG; Statistical Yearbook 1990, Metaleurop; Metalli Non Ferrosi 1991, Nuova SAMIM; Foreign trade data, OECD; Industrial Statistics Yearbook 1989, UN; Yearbook of Labour Statistics 1991, ILO.

Table 2.8 **Italy: production/consumption, trade and employment in the non-ferrous metals sector, average 1990/1991 ('000 metric tonnes)**

	Production	Imports	Exports	Consumption
Bauxite	4.5 [a]	1 673.2	9.6	
Alumina	778.5 [b]	203.3	400.9	
Aluminium	571.1	525.4 [c]	22.9 [c]	1 057.9
of which primary	224.8			661.0
of which secondary	346.3			396.9
Aluminium semi-manuf.[d]	710.2	235.9	179.3	
Copper ores/concs.		0.3 [e]	0.1 [a]	
Copper blister [f]		1.7	0.2	
Copper refined	83.2	409.3	19.7	472.8
Copper semi-manuf.	861.9	115.9	164.7	
Nickel refined				29.4
Lead ores/concs.	14.8	76.0 [g]	16.5 [g]	
Lead refined	189.3	92.1	19.6	258.5
of which secondary	99.4			
Zinc ores/concs.	39.7 [h]	406.2 [a]	11.0 [a]	
Zinc slab	252.1	74.1	43.0	276.5
Tin refined [i]		6.2	0.4 [g]	5.7
Employment in non-ferrous mining and metals ('000)		45 (1987)		
as a percentage of total labour force [j]		0.2%		
Exports of non-ferrous ores/metals (million US$)		1 709.6		
as a percentage of total exports		1.0%		
Imports of non-ferrous ores/metals (million US$)		4 585.3		
as a percentage of total imports		4.8%		

[a] Gross weight
[b] Alumina hydrate
[c] Including aluminium alloys
[d] Including alloy semis
[e] Copper content
[f] Including copper anodes
[g] 1989/1990
[h] Zinc content
[i] Tin content
[j] Total employment in metal ore mining and non-ferrous metals manufacturing (ISIC 230 and 372)

Source: World Metal Statistics, World Bureau of Metal Statistics; Metallstatistik 1980-1990, Metallgesellschaft AG; Statistical Yearbook 1990, Metaleurop; Metalli Non Ferrossi 1991, Nuova SAMIM; Foreign trade data, OECD; Industrial Statistics Yearbook 1989; Yearbook of Labour Statistics 1991, ILO.

Table 2.9 **Japan: production/consumption, trade and employment in the non-ferrous metals sector, average 1990/1991 ('000 metric tonnes)**

	Production		Imports		Exports		Consumption
Bauxite [a,k]			2 285.9		1.0		
Alumina	927.0	[b]	75.2	[c,k]	264.1	[c,k]	
Aluminium	1 493.1		2 741.3	[d]	4.7	[d]	4 093.9
of which primary	33.3						2 423.0
of which secondary	1 459.8						1 671.0
Aluminium semi-manuf. [e]	2 516.9		74.2		173.9		
Copper ores/concs.	12.7	[g]	3 621.5	[a]			
Copper blister [h]	1 063.0		30.3				
Copper refined	1 042.1		621.0		57.5		1 594.8
Copper semi-manuf. [f]	2 373.1		32.9		168.5		
Nickel ores/concs.			3 669.8	[a]			
Nickel refined	109.6		47.4		0.1		169.7
Lead ores/concs.	18.5	[i]	289.3	[a]			
Lead refined	329.7		97.2				419.3
of which secondary	117.2						
Zinc ores/concs.	130.1	[j]	1 162.3	[a]			
Zinc slab	709.1		141.0		19.7		829.9
Tin refined	0.8		33.6				34.9
Employment in non-ferrous mining and metals ('000)			119	(1988)			
as a percentage of total labour force			0.2%				
Exports of non-ferrous ores/metals (million US$)			2 106.9				
as a percentage of total exports			0.7%				
Imports of non-ferrous ores/metals (million US$)			11 088.9				
as a percentage of total imports			4.8%				

[a] Gross weight
[b] Alumina hydrate
[c] Including aluminium hydroxide
[d] Including aluminium alloys
[e] Including primary aluminium alloys
[f] Including alloy semis
[g] Copper content
[h] Including copper anodes
[i] Lead content
[j] Zinc content
[k] 1989/1990

Source: World Metal Statistics, World Bureau of Metal Statistics; Metallstatistik 1980-1990, Metallgesellschaft AG; Statistical Yearbook 1990, Metaleurop; Metalli Non Ferrossi 1991, Nuova SAMIM; Foreign trade data, OECD; Industrial Statistics Yearbook 1989, UN; Yearbook of Labour Statistics 1991, ILO.

Table 2.10 **Norway: production/consumption, trade and employment in the non-ferrous metals sector, average 1990/1991 ('000 metric tonnes)**

	Production		Imports		Exports		Consumption
Bauxite			11.8	[a]			
Aluminium	885.5		81.0		793.8		
of which primary	878.5						147.4
of which secondary	7.0						
Aluminium semi-manuf. [b]	176.3		27.6		144.3		
Copper ores/concs.	18.6	[c]	3.3	[d]	89.4	[d]	
Copper blister [e]	37.4						
Copper refined	35.8		0.3		37.7		8.0
Copper semi-manuf. [b]			19.7		3.0		
Nickel ores/concs. [f]	2.3						
Nickel refined	58.3				57.6		0.3
Lead ores/concs.	3.3						
Lead refined			4.1	[g]			3.8
Zinc ores/concs.	17.8		121.7	[a]	6.3	[a]	
Zinc slab	124.9		0.5		107.7		18.1
Tin refined	0.1		0.4				0.5
Employment in non-ferrous mining and metals ('000)			13.3 (1989)				
as a percentage of total labour force [h]			0.6%				
Exports of non-ferrous ores/metals (million US$)			2 654.5				
as a percentage of total exports			7.8%				
Imports of non-ferrous ores/metals (million US$)			1 477.4				
as a percentage of total imports			5.6%				

[a] 1989/1990
[b] Including alloy semis
[c] Copper content
[d] Gross weight
[e] Including copper anodes
[f] Nickel content
[g] Including lead alloys
[h] Total employment in metal ore mining and non-ferrous metals manufacturing (ISIC 230 and 372)

Source: World Metal Statistics, World Bureau of Metal Statistics; Metallstatistik 1980-1990, Metallgesellschaft AG; Statistical Yearbook 1990, Metaleurop; Metalli Non Ferrosi 1991, Nuova SAMIM; Foreign trade data, OECD; Industrial Statistics Yearbook 1989, UN; Yearbook of Labour Statistics 1991, ILO.

Table 2.11 **Portugal: production/consumption, trade and employment in the non-ferrous metals sector, average 1990/1991 ('000 metric tonnes)**

	Production	Imports	Exports	Consumption
Aluminium	2.0	49.8		49.0 [a]
of which secondary	2.0			
Copper ores/concs. [b]	158.5		176.5	
Copper refined	0.2	25.3		25.3
Copper semi-products [c]	39.1	23.7	7.6	
Nickel refined				0.2
Lead refined	4.5	18.5		23.0
of which secondary	4.5			
Zinc slab	3.8	10.9		10.2
Tin ores/concs.	2.2 [d]			
Tin refined	0.1			0.7
Employment in non-ferrous mining and metals ('000)		5.0 (1987)		
as a percentage of total labour force [e]		0.1%		
Exports of non-ferrous ores/metals (million US$)		316.2		
as a percentage of total exports [f]		2.2%		
Imports of non-ferrous ores/metals (million US$)		409.3		
as a percentage of total imports [f]		1.8%		

[a] Including waste uses
[b] Copper content
[c] Including alloy semis
[d] Tin content
[e] Total employment in metal ore mining and non-ferrous metals manufacturing (ISIC 230 and 372)
[f] 1989/1990

Source: World Metal Statistics, World Bureau of Metal Statistics; Metallstatistik 1980-1990, Metallgesellschaft AG; Statistical Yearbook 1990, Metaleurop; Metalli Non Ferrosi 1991, Nuova SAMIM; Foreign trade data, OECD; Industrial Statistics Yearbook 1989, UN; Yearbook of Labour Statistics 1991, ILO.

Table 2.12 **Spain: production/consumption, trade and employment in the non-ferrous metals sector, average 1990/1991 ('000 metric tonnes)**

	Production		Imports		Exports		Consumption
Bauxite			1 793.7	[a]	2.1	[a]	
Alumina	1 003.0		23.5	[a,b]	220.2	[a]	
Aluminium	446.7		46.6		112.7		
of which primary	355.3						292.5
of which secondary	91.4						
Aluminium semi-manuf.	277.6						
Copper ores/concs.	12.7	[c]	429.3	[d]	33.7	[a,d]	
Copper blister [e]	149.8		19.4		12.7		
Copper refined	180.2		24.3		45.5		151.1
Copper semi-manuf. [f]	207.0		123.1		35.0		
Nickel refined							17.9
Lead ores/concs.	55.4		50.3	[a]	61.7	[a]	
Lead refined	117.2		10.2		1.2		118.1
of which secondary	39.8						
Zinc ores/concs.	259.4		95.9	[a,d]	141.8	[a,d]	
Zinc slab	257.5		5.3		141.4	[g]	121.3
Tin ores/concs.	0.1						
Tin refined	1.5						3.4
Employment in non-ferrous mining and metals ('000)			27	(1988)			
as a percentage of total labour force			0.2%				
Exports of non-ferrous ores/metals (million US$)			987.4				
as a percentage of total exports			1.7%				
Imports of non-ferrous ores/metals (million US$)			1 535.5				
as a percentage of total imports			1.7%				

[a] 1989/1990
[b] Calcined alumina
[c] Copper content
[d] Gross weight
[e] Including copper anodes
[f] Including alloy semis
[g] Including zinc scrap

Source: World Metal Statistics, World Bureau of Metal Statistics; Metallstatistik 1980-1990, Metallgesellschaft AG; Statistical Yearbook 1990, Metaleurop; Metalli Non Ferrossi 1991, Nuova SAMIM; Foreign trade data, OECD; Industrial Statistics Yearbook 1989, UN; Yearbook of Labour Statistics 1991, ILO.

Table 2.13 **Sweden: production/consumption, trade and employment in the non-ferrous metals sector, average 1990/1991 ('000 metric tonnes)**

	Production	Imports	Exports	Consumption
Bauxite		52.7 [a]	4.0 [b]	
Alumina		236.5 [a]	0.3 [a]	
Aluminium	125.1			
of which primary	96.6	47.8 [c]	61.0 [c]	81.1
of which secondary	28.5			
Aluminium semi-manuf. [d]	95.6	105.1	47.5	
Copper ores/concs.	77.0 [e]	45.7 [f]	104.7 [f]	
Copper blister [g]	102.8	0.7	0.5	
Copper refined	96.9	61.4	35.8	115.2
Copper semi-manuf. [d]		35.4	92.0	
Nickel refined				17.9
Lead ores/concs. [h]	85.2		43.5 [a]	
Lead refined	82.1	1.1	58.2 [i]	25.3
of which secondary	39.8			
Zinc ores/concs. [j]	158.7		159.5 [a]	
Zinc slab		36.8	0.7 [k]	36.4
Tin refined				0.3
Employment in non-ferrous mining and metals ('000)		18.7 (1988)		
as a percentage of total labour force [l]		0.4%		
Exports of non-ferrous ores/metals (million US$)		1 157.0		
as a percentage of total exports		2.1%		
Imports of non-ferrous ores/metals (million US$)		1 120.5		
as a percentage of total imports		2.2%		

[a] 1989/1990
[b] 1989
[c] Including alloys
[d] Including alloy semis
[e] Copper content
[f] Gross weight
[g] Including copper anodes
[h] Lead content
[i] May include some bullion
[j] Zinc content
[k] 1990
[l] Total employment in metal ore mining and non-ferrous metals manufacturing (ISIC 230 and 372)

Source: World Metal Statistics, World Bureau of Metal Statistics; Metallstatistik 1980-1990, Metallgesellschaft AG; Statistical Yearbook 1990, Metaleurop; Metalli Non Ferrosi 1991, Nuova SAMIM; Foreign trade data, OECD; Industrial Statistics Yearbook 1989, UN; Yearbook of labour Statistics 1991, ILO.

Table 2.14 **United Kingdom: production/consumption, trade and employment in the non-ferrous metals sector, average 1990/1991 ('000 metric tonnes)**

	Production		Imports		Exports		Consumption
Bauxite			303.6		3.4	[a]	
Alumina	120.0		515.0	[b]	10.7	[a]	
Aluminium	420.3		219.0	[c]	143.4	[c]	496.1
of which primary	291.7						433.1
of which secondary	128.7						63.1
Aluminium semi-manuf.[d]	428.6	[e]	334.7		192.1		
Copper ores/concs.	0.6						
Copper blister [f]			47.8				31.9
Copper refined	95.9		246.9		13.7		299.3
Copper semi-manuf. [g]	423.8		141.9		113.4		
Nickel refined	27.6		17.1		14.9		31.1
Lead ores/concs. [h]	1.2		38.5				
Lead refined	320.2		34.3		76.2		282.6
of which secondary	160.1						
Zinc ores/concs.	3.8	[i]	69.3	[b]	17.7	[a,b]	
Zinc slab	97.0		121.9		11.6		188.4
Tin ores/concs.	2.9	[j]	16.0	[a,b]	1.7	[a,b]	
Tin refined	8.6		4.3		4.3		10.4
Employment in non-ferrous metals and mining ('000)			59	(1988)			
as a percentage of total labour force (k)			0.2%				
Exports of non-ferrous ores/metals (million US$)			2 214.4				
as a percentage of total exports			1.2%				
Imports of non-ferrous ores/metals (million US$)			4 180.6				
as a percentage of total imports			1.9%				

[a] 1989/1990
[b] Gross weight excluding re-exports
[c] Including primary alloys
[d] Including aluminium alloy semis
[e] Despatches of fabricated aluminium and aluminium alloys
[f] Including copper anodes
[g] Including copper alloy semis
[h] Lead content
[i] Zinc content
[j] Tin content
[k] Total employment in non-ferrous metals manufacturing (ISIC 372)

Sources: World Metal Statistics, World Bureau of Metal Statistics; Metallstatistik 1980-1990, Metallgesellschaft AG; Statistical Yearbook 1990, Metaleurop; Metalli Non Ferrossi 1991, Nuova SAMIM; Foreign trade data, OECD; Industrial Statistics Yearbook, 1989, UN; Yearbook of Labour Statistics 1991, ILO.

Table 2.15 **United States: production/consumption, trade and employment in the non-ferrous metals sector, average 1990/1991 ('000 metric tonnes)**

	Production		Imports		Exports		Consumption	
Bauxite [a]	272.5		13 712.1		67.3	[b]		
Alumina	5 422.5	[c]	4 245.5	[b]	1 255.6	[b]		
Aluminium	6 070.3						6 346.3	
of which primary	4 084.8		1 007.5		850.4		4 233.8	
of which secondary	1 985.5						2 112.5	[d]
Aluminium semi-manuf. [e]	5 261.5		410.9		520.0			
Copper ores/concs.	1 610.8	[f]	1 017.7	[g]	255.4	[g]		
Copper blister [h]	1 465.8		89.1		13.9			
Copper refined	1 986.2		291.2		242.0		2 104.1	
Copper semi-manuf. [i]	2 640.6		176.3		119.7			
Nickel refined	5.4		103.4		1.5		125.8	
Lead ores/concs. [j]	485.2		464.5		180.8			
Lead refined	1 242.5		90.0	[k]	50.6	[k]	1 260.8	
of which secondary	867.8							
Zinc ores/concs. [l]	544.9		171.0		434.9			
Zinc slab	367.9	[m]	589.9		2.2		962.0	
Tin ores/concs. [n]	0.1		1.3	[b]				
Tin refined	0.2		31.8		2.4		37.0	
Employment in non-ferrous mining and metals ('000)			251	(1989)				
as a percentage of total labour force [o]			0.2%					
Exports of non-ferrous ores/metals (million US$)			5 791.5					
as a percentage of total exports			1.5%					
Imports of non-ferrous ores/metals (million US$)			8 790.0					
as a percentage of total imports			1.7%					

[a] Gross weight
[b] 1989/1990
[c] Preliminary figure in 1991
[d] Including direct use of scrap
[e] Including alloy semis
[f] Copper content
[g] Excluding copper content of matte
[h] Including copper anodes
[i] Including alloy semis
[j] Lead content
[k] Including lead content of alloys
[l] Zinc content
[m] Including re-distilled zinc from primary and other smelters
[n] Tin content
[o] Total employment in non-ferrous metals manufacturing (ISIC 230 and 372)

Sources: World Metal Statistics, World Bureau of Metal Statistics; Metallstatistik 1980-1990, Metallgesellschaft AG; Statistical Yearbook 1990, Metaleurop; Metalli Non Ferrosi 1991, Nuova SAMIM; Foreign trade data, OECD; Industrial Statistics Yearbook 1989, UN; Yearbook of Labour Statistics 1991, ILO.

Table 3.1 **Australia: mining and processing facilities**

Name (location)	Type of operation	Company	Commodity
Mines			
Huntly (W. Australia)	OP	Alcoa of Australia	Al
Gove (Northern Territory)	OP	Nabalco Pty (Gove Aluminium & Swiss Aluminium Australia)	Al
Jarrahdale (W. Australia)	OP	Alcoa of Australia	Al
Weipa/Andoom (Queensland)	OP	Comalco	Al
Willowdale-Mt. William (W. Australia)	OP	Alcoa of Australia	Al
Worsley-Mt. Saddleback (W. Australia)	OP	Worsley Alumina Pty. (Reynolds & Billiton et al)	Al
Csa-Cobar (New South Wales)	UG	Enterprise Metals Pty. (CRA)	Cu, Ag, Pb, Zn
Mount Gunson (South Australia)	OP		Cu
Mount Isa (Queensland)	UG	Mount Isa Mines	Cu, Ag, Pb, Zn
Mount Lyell (Tasmania)	UG	Renison Gold Fields Consolidated	Cu, Ag, Au, Pyrites
Olympic Dam (South Australia)	UG	Roxby Mining Corp. (Western Mining Corp. & BP Australia)	Cu, Ag, Au, U
Greenvale (Queensland)	OP	Queensland Nickel Jt. Venture	Ni, Co
Kambalda Nickel (W. Australia)	UG	Western Mining Corp. Holdings	Ni
Leinster Nickel (W. Australia)	OP	Western Mining Corp. Holdings	Ni
Windarra (W. Australia)	UG	Western Mining Corp. Holdings	Ni
Greenbushes (W. Australia)	OP	Greenbushes	Sn, Li, Nb, Ta
Renison Bell (Tasmania)	UG	Renison Gold Fields Consolidated	Sn
Beltana (South Australia)	OP	Pasminco	Zn
Cadjebut (W. Australia)	UG	BHP-Utah Metals Minerals/ Billiton Aust.	Zn, Pb
Elura (New South Wales)	UG	Pasminco	Zn, Ag, Au
Hellyer-Cleveland (Tasmania)	UG	Aberfoyle	Zn, Ag, Pb
Hilton (Queensland)	UG	Mount Isa Mines	Zn, Ag, Pb
Broken Hill North & South (New South Wales)	OP/UG	Pasminco	Zn, Ag, Pb
Que River (Tasmania)	UG	Aberfoyle	Zn, Ag, Au
Rosebery (Tasmania)	UG	Pasminco	Zn, Ag, Au, Cu, Pb
Scuddles-Golden Grove (W. Australia)	UG	Murchison Zinc Co. Ph.	Zn, Ag, Au, Cu
Thalanga (Queensland)	OP	Pancontinental Mining (Outokumpu & Agip)	Zn, Ag, Au, Cu, Pb
Woodcutters (Northern Territory)	UG	Nicron Resources	Zn, Ag, Pb
Woodlawn (New South Wales)	UG	Denehurst/CRA	Zn, Ag, Au, Cu, Pb
Zinc Corporation (New South Wales)	UG	Pasminco	Zn, Ag, Pb
Smelters/Refineries			
Kwinana (W. Australia)	Ref	Alcoa of Australia	Al
Pinjarra (W. Australia)	Ref	Alcoa of Australia	Al
Wagerup (W. Australia)	Ref	Alcoa of Australia	Al
Gladstone (Queensland)	Ref	Queensland Alumina (Kaiser, Howmet, Alcan, Comalco)	Al
Gove (Northern Territory)	Ref	Gove Joint Venture (Gove Aluminium)	Al
Worsley (W. Australia)	Ref	Worsley Alumina Pty (Billiton & Reynolds)	Al

Table 3.1 (cont'd) **Australia: mining and processing facilities**

Name (location)	Type of operation	Company	Commodity
Bell Bay (Tasmania)	Sme	Comalco	Al
Point Henry, Geelong (Victoria)	Sme	Alcoa of Australia	Al
Kurri-Kurri (New South Wales)	Sme	Alcan Australia	Al
Boyne Island (Queensland)	Sme	Boyne Smelters	Al
Tomago (New South Wales)	Sme	Tomago Aluminium Pty	Al
Portland (Victoria)	Sme	Portland Smelter Joint Venture [Aluvic (Govt. of Victoria), Alcoa et al]	Al
Southern Copper (New South Wales)	Sme-Ref	Electrolytic Refining & Smelting Co. of Australia	Cu
Mount Isa (Queensland)	Sme-Ref	Mount Isa Mines	Cu, Pb, Zn
Townsville (Queensland)	Ref	Copper Refineries Pty (MIM Holdings)	Cu
Olympic Dam (South Australia)	Sme-Ref	Western Mining	Ni
Kalgoorlie (W. Australia)	Sme	Western Mining	Ni
Kwinana (W. Australia)	Ref	Western Mining	Ni
Yabalu (Queensland)	Ammonia leaching	Greenvale (Metals Exploration & Freeport Queensland Nickel)	Ni
Port Pirie (South Australia)	Sme-Ref	Pasminco/Broken Hill Assoc. Smelter Pty	Pb, Zn
Cockle Creek (New South Wales)	Sme	Pasminco/Sulphide Corp. Pty	Pb
Mount Isa (Queensland)	Sme	Mount Isa Mines	Pb
Risdon (Tasmania)	Ref	Pasminco/Electrolytic Zinc Co. of Austr.	Zn

UG = Underground Mine; OP = Open Pit Mine; Sme = Smelter; Ref = Refinery
Source: Mining Magazine; US Bureau of Mines, ADIT; Metaleurop, Annuaire Statistique; Metal Bulletin; ILZSG

Table 3.2 **Belgium: processing facilities**

Location	Type of operation	Company	Commodity	Capacity ('000t)
Smelters-Refineries				
Olen	Refinery	MHO (Acec-Union Minière)	Cu	300
Beerse	Smelter-Refinery	Metallo-Chimique N.V.	Cu	190
Hoboken	Smelter-Refinery	MHO (Acec-Union Minière)	Pb	125
Overpelt[1]	Smelter-Refinery	Vieille-Montagne (Acec-Union Minière)	Zn	25
Balen-Wezel	Smelter-Refinery	Vieille-Montagne (Acec-Union Minière)	Zn	190
Secondary plants				
Tienen		Affilips	Al,Cu,Pb,Zn	
Bruxelles		Johnson Matthey	Ag,Au,Pt	
Beerse		Chemisch en Metallurgisch Bedrijf Campine	Pb	
Bruxelles		Fonderie et Manufacture de Métaux	Pb	
Zolder		Rezinal	Zn	
Semi-manufactures				
Duffel	Sheets;strips	Hoogovens Aluminium	Al	
Liège	Foils	Phenix Aluminium (Hoogovens)	Al	
Seneffe	Extrusions	Hydalu (Norsk Hydro)	Al	
Raeren	Extrusions	Hydalu (Norsk Hydro)	Al	
Ghlin	Rods;wire;sheets; strips;foils	Aleurope (Reynolds)	Al	
Lichtervelde	Extrusions	Remi Claeys Aluminium	Al	
Hemiksem	Cables;wires; sections	Lamitref Metals	Al, Cu	
Liège	Sheets;strips;tubes	Boliden Cuivre et Zinc (Trelleborg)	Cu	
Liège	Leaf;strips; plates;sheets	Uca	Cu	
Bruxelles	Semis;salts;nitrates	Johnson Matthey	Ag, Au, Pt	
Overpelt	Semis	Vieille-Montagne (Acec-Union Minière)	Pb	
Vilvoorde	Alloys;semis	N.V. Stoop	Pb	
Angleur	Alloys;semis; powders	Vieille-Montagne (Acec-Union Minière)	Zn	

1. The Overpelt electrolytic plant of 120 000 tonnes was closed on 1 January 1992; refinery capacity of 25 000 tonnes remains in operation.

Source: Fédération des Entreprises de Métaux non ferreux

Table 3.3 **Canada: mining and processing facilities**

Province/company	Commodity	Production 1990 (Tonnes) (Metal contained)	
Mines			
British Columbia			
Noranda Minerals Inc.	Cu	21 300	
Bell Mine Equity Silver Mines Limited	Cu	6 250	
Gilbraltar Mines Limited	Cu	32 500	
Highland Valley Copper	Cu	163 700	
Afton Mines Ltd.	Cu	10 900	
Newmont Mines Limited	Cu	26 000	
Princeton Mining Corp.	Cu	25 600	
Brenda Mines Ltd.	Cu	6 200	
BHP - Utah Mines Ltd.	Cu	44 850	
Minnova Inc.	Cu	1 900	
Westmin Resources Limited	Cu	21 050	
Manitoba			
HBMS, Ruttan mine	Cu	19 100	
INCO Limited, Thompson	Cu	3 500	
HBMS, Flin Flon	Cu	33 500	
HBMS/Outokumpu JV	Cu	3 150	
Ontario			
Noranda Minerals Inc., Lyon Lake	Cu	3 150	
Noranda Minerals Inc., Geco mine	Cu	21 050	
Falconbridge Limited, Timmins	Cu	122 400	
Minnova Inc.	Cu	3 550	
Falconbridge Limited, Sudbury	Cu	33 650	
INCO Limited, Sudbury	Cu	103 400	
Quebec			
BP Canada Inc. (Les Mines Selbaie)	Cu	28 500	
Noranda Minerals Inc., Mattagami Division	Cu	3 400	
Noranda Minerals Inc., (Isle Dieu and Norita)	Cu	3 400	
Minnova Inc., Lac DuFault Division	Cu	33 900	
Westminer Canada Limited	Cu	2 900	
Noranda Minerals Inc., Mine Gaspe	Cu	20 650	
Minova Inc., Opemiska Division	Cu	5 850	
Campbell Resources Inc.	Cu	850	
Lac Minerals Ltd.	Cu	2 350	
Breakwater Resources Ltd./Brookline Minerals Inc. JV	Cu	480	
New Brunswick			
Brunswick Mining and Smelting Corporation Limited	Cu	6 600	
Breakwater Resources	Cu	900	
Noranda Minerals Inc. (Heath Steele & Stratmat)	Cu	2 350	
Nova Scotia			
Rio Kemptville Tin Corporation	Cu	480	
		Zinc	**Lead**
Newfoundland			
Newfoundland Zinc Mines	Zn/Pb	17 080	52 300
Nova Scotia			
Westminer Canada Limited	Zn/Pb	8 360	4 300
New Brunswick			
Brunswick Mining and Smelting Corporation Limited	Zn/Pb	172 200	52 300
Breakwater Resources	Zn/Pb	26 300	6 700
Noranda Inc.	Zn/Pb	50 280	13 500

Table 3.3(cont'd) **Canada: mining and processing facilities**

Province/company	Commodity	Production 1990 (Tonnes) (Metal Contained)	
		Zinc	**Lead**
Quebec			
Abcourt Mines Inc.	Zn/Pb	2 000	—
BP Resources Canada Limited	Zn/Pb	49 400	2 100
Breakwater Resources Brookline Minerals Inc., JV	Zn/Pb	840	200
Deak Resources Corp.	Zn/Pb	890	—
Noranda Inc., Mattagami	Zn/Pb	53 500	760
Normetal Mining & Exploration Inc.	Zn/Pb	3 400	—
Audrey Resources Inc.	Zn/Pb	8 500	—
Ontario			
Falconbridge Limited, Timmins	Zn/Pb	177 250	2 950
Noranda Inc., Lyon Lake	Zn/Pb	27 000	3 400
Noranda Inc., Geco	Zn/Pb	45 650	—
Minnova Inc., Winston Lake	Zn/Pb	58 900	—
Manitoba			
HBMS, Flin Flon	Zn/Pb	76 400	3 300
HBMS, Ruttan mine	Zn/Pb	19 200	—
British Columbia			
Minnova Inc., Somotasum Mine	Zn/Pb	4 400	2 300
Cominco Ltd., Sullivan Mine	Zn/Pb	20 500	15 300
Treminco Resources Ltd.	Zn/Pb	1 780	1 700
Westmin Resources Limited	Zn/Pb	37 900	2 000
Yukon			
Curragh Resources Corporation	Zn/Pb	191 250	118 000
Northwest Territories			
Cominco Inc., Polaris mine	Zn/Pb	143 000	39 000
Nanisivik Mines Limited	Zn/Pb	51 000	2 000
Ontario			
Falconbridge Limited, Sudbury	Ni	34 400	
INCO Limited, Sudbury	Ni	102 000	
INCO Limited, Shebandowan mine	Ni	n.a.	
Timmins Nickel Inc.	Ni	2 300	
Manitoba			
INCO Limited, Thompson	Ni	52 000	
Hudson Bay Mining and Smelting Co. (HBMS), Namew Lake	Ni	8 000	
Nova Scotia			
Rio Kemptville Tin Corporation	Sn	n.a.	

Table 3.3(cont'd) **Canada: mining and processing facilities**

Province/company	Type of operation	Commodity	Production 1990 (Tonnes) (Metal Contained)
Smelters/Refineries			
Quebec			
Alcan Aluminium Limited			
Grande Baie	Smelter	Al	171 000
Jonquière	Smelter	Al	282 000
Isle-Maligne	Smelter	Al	73 000
Shawinigan	Smelter	Al	84 000
Beauharnois	Smelter	Al	47 000
Laterrière	Smelter	Al	150 000
Canadian Reynolds Metal			
Baie Comeau	Smelter	Al	280 000
Aluminerie de Bécancour Inc.			
Bécancour	Smelter	Al	280 000
British Columbia			
Alcan Aluminium Limited			
Kitimat	Smelter	Al	268 000
			(Rated capacity) (Tonnes of ores and concentrates)
Manitoba			
HBMS, Flin Flon (copper anodes)	Smelter	Cu	320 000
Ontario			
Falconbridge Limited, Timmins (molten "blister" copper)	Smelter	Cu	440 000
Falconbridge Limited, Sudbury (copper nickel matte)	Smelter	Cu	600 000
INCO Limited, Sudbury (blister copper)	Smelter	Cu	500 000
Quebec			
Noranda Inc., Noranda (copper anodes)	Smelter	Cu	770 000
Noranda Inc., Murdochville (copper anodes)	Smelter	Cu	221 000
			(Rated capacity) (concentrate and copper scrap)
Ontario			
Falconbridge Limited, Timmins	Refinery	Cu	95 000
INCO Limited, Sudbury	Refinery	Cu	170 000
Quebec			
Noranda Inc., Montreal	Refinery	Cu	350 000
British Columbia			
Gibraltar Mines Limited	Refinery	Cu	5 500

Table 3.3(cont'd) **Canada: mining and processing facilities**

Province/company	Type of operation	Commodity	Production 1990 (Tonnes) (Metal Contained)
Smelters/Refineries (cont'd)			
Ontario			
INCO Limited, Sudbury	Smelter	Ni	110 000
Falconbridge Limited, Sudbury	Smelter	Ni	45 000
Manitoba			
INCO Limited, Thompson	Smelter	Ni	81 600
Ontario			
INCO Limited, Sudbury	Refinery	Ni	56 700
Manitoba			
INCO Limited, Thompson	Refinery	Ni	49 900
Alberta			
Sherritt Gordon Limited Fort Saskatchewan	Refinery	Ni	25 000
New Brunswick			
Brunswick Mining and Smelting Corporation Limited	Smelter	Pb	72 000
British Columbia			
Cominco Inc., Trail	Smelter	Pb	135 000
Ontario			
Tonolli Canada Limited	Secondary smelter	Pb	35 000
The Canada Metal Company	Secondary smelter	Pb	12 000
Quebec			
Nova Pb Inc.	Secondary smelter	Pb	60 000
Manitoba			
The Canada Metal Company	Secondary smelter	Pb	5 000
British Columbia			
Metalex Products Ltd.	Secondary smelter	Pb	5 000
Quebec			
Noranda Inc., Valleyfield	Smelter	Zn	230 000
Ontario			
Falconbridge Limited, Timmins	Smelter	Zn	133 000
Manitoba			
HBMS, Flin Flon	Smelter	Zn	82 000
British Columbia			
Cominco Inc., Trail	Smelter	Zn	300 000

Source: Energy, Mines and Resources Canada.

Table 3.4 **France: Mining and Processing Facilities**

Name/location	Type of operation	Company	Commodity	Capacity ('000t)
Mines				
Salsigne (Aude)	UG	Metaleurop	Cu,Au,Ag	1
Noilhac Saint Salvy (Tarn)	UG	Metaleurop	Zn,Ag	30
CGMC (Moneo) (New Caledonia)	OP	Ballande	Ni	400
Group Pentecost (New Caledonia)	OP		Ni	1 000
Kouaoua (New Caledonia)	OP	SLN	Ni	1 200
Montegnat (New Caledonia)	OP		Ni	900
Nepoui (New Caledonia)	OP	SLN	Ni	300
Poro (New Caledonia)	OP	SLN	Ni	
Thio (New Caledonia)	OP	SLN	Ni	1 000
Smelters/Refineries				
Gardanne (Bouches-du-Rhône)	Ref	Péchiney	Al	700
Auzat (Ariège)	Sme	Péchiney	Al	44
Lannemezan (Htes-Pyrénées)	Sme	Péchiney	Al	44
St. Jean de Maurienne (Savoie)	Sme	Péchiney	Al	125
Venthon (Savoie)	Sme	Péchiney	Al	31
Dunkerque (Nord)	Sme	Péchiney	Al	215
Poissy (Yvelines)	Sme	Sté Francaise d'affinage du cuivre	Cu	10
Le Palais (Hte Vienne)	Sme	Cie Gle d'électrolyse	Cu	45
Toulouse (Hte Garonne)	Sme	Raffinerie du Sud-Ouest	Cu	
Sandouville-Le Havre	Sme	Eramet-SLN	Ni	16
Noyelles-Godault (Pas-de-Calais)	Sme-Ref	Metaleurop	Pb,Zn	260
Auby-les-Douai (Nord)	Sme-Ref	Vieille Montagne	Zn	115
Doniambo (New Caledonia)	FeNi Matte Sme	SLN	Ni	62
Secondary plants				
Villefranche-sur-Saône		Metaleurop	Pb	40
Escadoeuvres (Nord)		Metaleurop	Pb	27
Toulouse (Hte-Garonne)		Sté de traitement chimique des métaux	Pb	25
Bazoches-les-Gallerandes		Sté de traitement chimique des métaux	Pb	30
Bourg Fidèle (Ardennes)		Métal Blanc	Pb	13
Pont-Sainte-Maxence (Oise)		Sté G.A.S.T.	Pb	20
L'isle Jourdain		Aldevienne	Zn	2
Metz		Sté Nelle des fonderies zinc	Zn	2
Fécamp		Mapral	Zn	10
La Plaine St. Denis		Georges Ostrowiak S.A.	Zn	10
Paris		Ramb	Zn	2
Lunéville		Sté Nelle des métaux	Zn	2

UG = Underground Mine; OP = Open Pit Mine; Sme = Smelter; Ref = Refinery.
Source: Metaleurop, Annuaire Statistique; Mining Magazine; ILZSG

Table 3.5.1 **Germany: Western Germany — mining and processing facilities**

Location	Type of operation	Company	Commodity	Capacity ('000t)
Mines				
Bad Grund/Harz (closed in 1992)	UG	Preussag AG Metall	Zn, Pb	
Meggen, Lennestadt (closed in 1991)	UG	Sachtleben Bergbau GmbH	Pyrites, Zn, Pb	
Smelters/Refineries				
Rheinfelden	Sme	Aluminium Rheinfelden GmbH	Al	42
Stade	Sme	Aluminium Oxid Stade GmbH (VAW)	Al	70
Hamburg	Sme	Hamburger Aluminium Werk GmbH	Al	120
Voerde	Sme	Hoogovens Aluminium GmbH	Al	97
Essen	Sme	Leichtmetall GmbH	Al	
Bergheim	Sme	Martinswerk GmbH	Al	
Töging, Innwerk	Sme	Vereinigte Aluminium-Werke AG	Al	87
Lünen, Lippewerk	Ref-sme	Vereinigte Aluminium-Werke AG	Al	
Stüttgen, Rheinwerk	Sme	Vereinigte Aluminium-Werke AG	Al	210
Grevenbroich, Erftwerk	Sme	Vereinigte Aluminium-Werke AG	Al	
Pocking, Rottwerk	Sme	Vereinigte Aluminium-Werke AG	Al	
Lünen	Ref	Hüttenwerke Kayser AG	Cu	115
Hamburg	Sme-	Norddeutsche Affinerie AG	Cu	150
	Ref		Cu	290
Binsfeldhammer	Sme-ref	Berzelius Metallhütten GmbH	Pb	90
Duisburg	Sme	Berzelius Metallhütten GmbH	Pb	40
Hamburg	Sme	Norddeutsche Affinerie AG	Pb	50
Nordenham	Sme-ref	Metaleurop Weser Blei GmbH	Pb	90
Datteln	Ref	Ruhr-Zink GmbH	Zn	200
Nordenham	Sme	Metaleurop Weser Zink GmbH	Zn	130
Duisburg	Sme-Ref	Berzelius Metallhütten GmbH	Zn	100
Harlingerode	Sme	Metaleurop Harz Metall Gmbh	Zn	30

UG = Underground mine; Sme = Smelter; Ref = Refinery
Source: Yearbook of Mining, Oil and Gas, Electricity, Chemicals; ILZSG; Nuova SAMIM, Metalli Non Ferrosi; Metaleurop, Annuaire statistique

Table 3.5.2 **Germany: New Federal Länder — mining and processing facilities**

Location	Company	Commodity
Mines		
Mansfeld Niederröblingen (closed September 1990)		Cu, Ag
Altenberg (closed March 1991)		Sn
Ehrenfriedersdorf (closed September 1990)		Sn
Smelters/Refineries		
Lutherstadt Eisleben	Mansfeld AG	Cu, Al
Aue	Aue Nickel Smelter	Ni, Cu
Freiberg	Saxonia AG	Pb, Zn, Cu

Source: Yearbook of Mining, Oil and Gas, Electricity, Chemicals.

Table 3.6 **Greece: mining and processing facilities**

Name/location	Type of operation	Company	Commodity
Mines			
Itea-Distomon-Amphissa (Phokis)	OP-UG	Hellenic Bauxites of Distomon + Delphi-Distomon (Aluminium de Grèce)	Al
Eleusis;Itea;Kimi(Euboea); Oetea(Lamia)	OP-UG	Mining Industrial & Shipping/ (Eleusis Bauxite Mines-Elbaumin)	Al
Elikon-Levadia (Boetia)	UG	Greek Helikon Bauxites (G L Barlos)	Al
Parnasse-Ghiona/ Itea-Euboea	OP-UG	Bauxite Parnasse Mining Co.	Al
Eretria (Volos)	OP	Financial Mining Industrial & Shipping Corp.-Fimisco-Elbaumin	Cr
Vourinos Area [Aetoraches/ Koursoumia/Rizo](Kozani)	UG OP	Hellenic Ferroalloys-Elsi (Himic)	Cr
Xerolivado-Skoumtsa (Macedonia)	UG	Hellenic Ferroalloys-Elsi (Himic)	Cr
Nevrokopi (Drama)	UG	Elbaumin	Mn
Ayios Ioannis (Larymna)	OP-UG	Hellenic Mining & Metallurgical Co. of Larymna - Larco	Ni, Cr
Pagonda-Vrissakia (Euboea)			
Laurion (Attica)		Emmel	Pb
Madem Lakkos-Mavres Petres (Chalkidiki)	OP-UG	Kassandra Mines (Hellenic Chemical Prods & Fertilisers)	Zn, Pb, Fe
Olympias (Chalkidiki)			Zn, Pb
Molai (Laconia)	under development	Aegean Metallurgical Industries (Himic)	
Smelters/Refineries			
Aspra Spitia (Boetia)	Ref	Aluminium de Grèce	Al
St. Nicholas-Distomon (Boetia)	Sme	Aluminium de Grèce	Al
Kamiotissa (Itea)	Ref	Bauxites Parnasse Mining	Al
Aghia Efthymia (Thisvi)	Ref (under develop-ment)	Hellenic Alumina (Hellenic Industrial Development Bank-Etba + Tsvetmetexport)	Al
Lavrion (Attika)	Sme-Ref	Greek Metallurgical and Mining Co. of Lavrion SA (Emmel)	Pb
Larymna (Lokris)	Sme	Larco	FeNi
Politika (Euboea)	Sme	Larco	FeNi
Tsigeli (Volos)	Sme	Hellenic Ferroalloys (Himic)	FeCr

UG = Underground Mine; OP = Open Pit Mine; Sme = Smelter; Ref = Refinery
Source: Ministry of Industry, Energy and Technology.

Table 3.7 **Ireland: mining and processing facilities**

Name/location	Type of operation	Company	Commodity
Mines			
Mavan (Meath)	Underground	Tara Mines (Outokumpu)	Zn, Pb
Galmoy	Underground (under construction)	Conroy Petroleum	Zn, Pb
Refineries			
Askeaton (Limerick)	Alumina refinery	Aughinish (Alcan & Billiton)	Al
Secondary plants			
Kells (Meath)		P. Carney Ltd.	Al
Dublin (Meath)		Metal Refiners Ltd.	Pb, Cu
Dublin (Meath)		Mining Co. of Ireland	Pb

Source: Metal Bulletin Metaleurop, Annnuaire Statistique, ILZSG.

Table 3.8 **Italy: mining and processing facilities**

Name/location	Type of operation	Company	Commodity	Production ('000t)
Mines				
Monteponi (Sardinia)	UG		Pb, Zn	440
Masua (Sardinia)	UG		Pb, Zn	470
Smelters/Refineries				
Fusina (Venezia)	Sme	Aluminix	Al	39
Porto Vesme (Cagliari)	Sme	Aluminix	Al	197
Porto Marghera (Venezia)	Ref	Nuova SAMIM	Cu	60
Pieve Vergonte (Novara)	Ref	Nuova SAMIM	Cu	22
Fornaci Di Barga (Lucca)	Ref	Europametalli - LMI	Cu	24
San Gavino (Cagliari)	Sme	Nuova SAMIM	Pb	105
Porto Vesme (Cagliari)	Sme	Nuova SAMIM	Pb	30
Porto Vesme (Cagliari)	Sme/Ref	Nuova SAMIM	Zn	70
Crotone (Catanzaro)	Ref	Pertusola Sud	Zn	100
Secondary plants				
Paderno Dugnano (Milano)		Nuova SAMIM	Pb	50
Marcianise (Caserta)		Nuova SAMIM	Pb	40
Arcola (La Spezia)		Metalli & Derivati	Pb	15
Brugherio (Milano)		Piomboleghe	Pb	15
Borgo San Siro (Pavia)		Sarpi	Pb	
Maclodio (Brescia)		Piombifera Bresciana	Pb	20
Burago Molgora		Capitelli	Zn	
Brugherio		Ilcomet	Zn	

UG = Underground Mine; Sme = Smelter; Ref = Refinery

Source: Nuova SAMIM, Metalli Non Ferrosi; Mining Magazine; International Lead and Zinc Study Group.

Table 3.9 **Japan: mining and processing facilities**

Name (location)	Type of operation	Company	Commodity
Mines			
Kamaishi (Iwate)	UG	Kamaishi Mining Co. (Nyttetsu)	Cu
Uchinotai (Kosaka, Akita)	UG	Uchinotai Mining Co. Ltd. (Dowa)	Pb, Zn
Hanaoka (Akita)	UG	Hanaoka Mining Co. (Dowa)	Pb, Zn
Nurukawa (Aomori)	UG	Uchinotai Mining Co. Ltd. (Dowa)	Pb, Zn
Kamioka (Gifu)	UG	Kamioka Mng. & Smltg. Co. (Mitsui Metal)	Pb,Zn,Ag
Toyoha (Hokkaido)	UG	Toyoha Mining Co. (Nippon Mining)	Pb,Zn,Ag
Other Mines			
Kushikino (Kagoshima)	UG	Mistui Kushikino Mining Co. Ltd. (Mitsui)	Au,Ag
Hishikari (Kagoshima)	UG	Sumitomo Metal & Mining Co. Ltd.	Au,Ag
Smelters/Refineries			
Hitachi (Ibaragi)	Ref	Nippon Mining Co. Ltd.	Cu
Bessi-Niihama (Ehime)	Ref	Sumitomo Metal & Mining Co. Ltd.	Cu,Ni
Kosaka (Akita)	Sme-Ref	Kosaka Smelting Co. Ltd. (Dowa)	Cu,Pb
Naoshima (Kagawa)	Sme-Ref	Mitsubishi Metal Corp.	Cu
Onahama (Fukushima)	Sme-Ref	Onahama Smltg (Mitsubishi/Dowa/Furukawa)	Cu
Saganoseki (Ohita)	Sme-Ref	Nippon Mining Co. (Mistubishi/Furukawa/Nittetsu)	Cu,Pb
Tamano (Okayama)	Sme-Ref	Hibi Kyodo Smelting Co. (Mitsui/Furukawa)	Cu
Naoshima (Kagawa)	Sme-Ref	Mitsubishi/Cominco	Pb
Takehara (Hiroshima)	Sme-Ref	Mitsui Mining & Smelting	Pb,Sn,Cu
Other Smelters/Refineries			
Chigirishima (Hiroshima)	Sme-Ref	Toho Zinc Co.	Pb, Zn
Hachinohe (Aomori)	Sme-Ref	Hachinohe Smelting Co. (Mitsui/Nippon/Dowa/...)	Pb, Zn
Harima (Hyogo)	Sme-Ref	Sumitomo Metal Mining Co. Ltd.	Pb, Zn
Hosokura (Miyagi)	Sme-Ref	Hosokura Smelting Co. (Mitsubishi Metal)	Pb, Zn
Kamioka (Gifu)	Sme-Ref	Kamioka Mng & Smltg Co. (Mitsui)	Pb, Zn
Iijima (Akita)	Sme-Ref	Akita Zinc Co. (Dowa/Nippon/Sumitomo/...)	Zn
Mikkaichi (Toyama)	Sme-Ref	Nikko Zinc Co. (Nippon Mining)	Zn
Akita (Akita)	Sme-Ref	Mitsubishi Metal Corp.	Zn
Annaka (Gunma)	Sme-Ref	Toho Zinc Co. Ltd.	Zn
Hikoshima (Yamaguchi)	Sme-Ref	Hikoshima Smelting Co. (Mitsui)	Zn
Kambara (Shizuoka)	Sme	Nippon Light Metal Co.	Al
Toyo (Ehime)	Sme-Ref	Sumitomo Metal & Mining Co.	Cu
Oeyama (Kyoto)	Ref	Nippon Yakin Kogyo Co.	Ni
Hachinohe (Aomori)	Sme	Pacific Metals Co.	Ni, Fe
Hyuga (Miyasaki)	Sme	Sumitomo Metal Mining Co.	Ni
Niihama (Ehime)	Ref	Sumitomo Metal Mining Co.	Ni
Matsuzaka (Mie)	Ref	Tokyo Nickel	Ni
Secondary plants			
Kita Kyushu-Shi		Daishin Sagyo Co. Ltd	Pb
Ichikawa-Shi		Ichikawa Ref Co.	Pb
Yawata-Shi		Nippon Kappan Jigane Co. Ltd	Pb
Amagasaki-Shi		Osaka Lead Ref Co. Ltd	Pb
Utsunomiya-Shi		Tamura Sangyo Co.	Pb
Kariya-Shi		Tomatsu Metallurgical Co.	Pb
Iwatsuki-Shi		Eiwa Kinzoku Co. Ltd	Pb
Osaka		Hayari Metal Industrial Co.	Zn
Osaka		Osaka Zinc Industrial Co.	Zn
Tokyo		Toyo Zinc Co.	Zn

UG = Underground Mine; Sme = Smelter; Ref = Refinery
Source: MITI, Collection of Mineral Information; Metaleurop Annuaire Statistique; Mining Magazine; Nuova SAMIM, Metalli Non Ferrosi; International Lead and Zinc Study Group

Table 3.10 **Norway: mining and processing facilities**

Name (location)	Type of operation	Company	Commodity	Production 1989 (Tonnes)
Mines				
Bidjovagge (Kautokeino)	OP	A/S Bidjovagge Gruber (Outokumpu)	Cu	24 888
Grong (Røyrvik, Nord-Trondelag)	UG	Grong Gruber A/S (Outokumpu/Borregaard)	Cu	25 274
			Zn	12 885
Tverrfjellet (Hjerkinn)	UG	Folldal Verk A/S (Outokumpu/Borregaard)	Cu	28 023
			Zn	5 180
			Pyrites	200 366
Ballangen (Naroil-Pana)	UG	Nikkel & Olivin A/S	Cu, Ni	
Haug I Dalane	OP	Titania A/S (NL Industries)	Ni	460
			Cu	170
			Ilmenite	930 000
Bleikvassli (Nordland)	UG	Bleikvassli Gruber A/S (A/S Sydvaranger)	Cu	184
			Zn	13 091
Smelters				
Årdal	Sme	Hydro Aluminium A/S (Norsk Hydro)	Al	172 000
Karmøy	Sme	Hydro Aluminium A/S (Norsk Hydro)	Al	220 000
Sunndalsøra	Sme	Hydro Aluminium A/S (Norsk Hydro)	Al	137 000
Husnes	Sme	Sør-Norge Aluminium A/S (Norsk Hydro & Alusuisse)	Al	68 000
Mosjøen	Sme	Elkem Aluminium (Elkem & Alcoa)	Al	115 000
Lista	Sme	Elkem Aluminium (Elkem & Alcoa)	Al	90 000
Vennesla	Sme	Vigeland Metal Refinery	Al	5 000
Kristiansand	Sme	Falconbridge Nikkelverk A/S (Boliden & Noranda)	Ni	55 000
			Cu	35 000
Odda	Sme	Norzink A/S (Boliden & RTZ)	Zn	125 000

UG = Underground mine; OP = Open pit mine; Sme = Smelter.
Source: Norwegian Royal Ministry of Industry

Table 3.11 **Portugal: mining and processing facilities**

Name (location)	Type of operation	Company	Commodity
Mines			
Aljustrel (Alentejo)	Underground	Empresa de Desenvolvimento Mineiro (EDM)	Cu,Pb,Zn
Lousal/Caveira	Underground		Cu, pyrites
Neves Corvo (Alentejo)	Underground	Sociedade Mineira de Neves-Corvo SA	Cu,Zn,Sn
Panasqueira (Barroca Grande)	Underground	Beralt	W,Sn,Cu
Smelters-Refineries			
Barreiro	Smelter-Refinery	Quimigal E. P.	Cu,Zn
Secondary plants			
Azambuja		Sonalur (Tudor)	Pb
Lisboa		Autosil	Pb
Matosinhos		Fonseca & Flora	Zn
Oeiras		Autosil	Zn

Source: Metaleurop, Annuaire Statistique; Mining Magazine; Ministère de l'Industrie et de l'Énergie du Portugal

Table 3.12 **Spain: mining and processing facilities**

Name (location)	Type of operation	Company	Commodity
Mines			
Sotiel (Huelva)	UG	Minas de Almagrera	Cu, Pb, Zn
La Zarza (Huelva)	UG	Minas de Tharsis	Cu,pyrites
Tharsis (Huelva)	OP	Minas de Tharsis	Cu,pyrites
Cartagena (Murcia)	Surface	Peñarroya (Metaleurop)	Pb, Zn, Ag
Linares (Jaen)	UG	Minas de la Cruz	Pb
Aznalcollar (Sevilla)	OP	Andaluza de Piritas (Boliden)	Zn, Cu, Pb, Ag
Reocin (Santander)	UG	Asturiana de Zinc	Zn, Pb
La Troya (Guipuzcoa)	UG	Exminesa	Zn, Pb
Rubiales (Lugo)	UG	Exminesa (Cominco)	Zn, Pb, Ag
Smelters/Refineries			
San Ciprian	Ref	Inespal	Al
Aviles (Asturias)	Sme	Inespal	Al
La Coruña	Sme	Inespal	Al
Huelva	Sme-Ref	Rio Tinto Minera, S.A.	Cu
Barcelona	Ref	Electrolisis del Cobre	Cu
Asua - Berango (Vizcaya)	Sme-Ref	Empresas Reunidas de Cobre Electrolitico Y Metales-Ercosa	Cu
Cartagena (Murcia)	Sme	Peñarroya (Metaleurop)	Pb
Cartagena (Murcia)	Ref	Española del Zinc	Zn
San Juan De Nieva (Aviles Asturias)	Ref	Asturiana de Zinc	Zn
Erandio (Bilbao)	Ref	Metalquimica del Nervión	Zn
Secondary plants			
Erandio (Vizcaya)		Remetal	Al
Valladolid		Refinalsa	Al
Barcelona		Productos Agma	Al
Leioa (Vizcaya)		Metales Iberica Aranzadi	Cu
Cubas de la Sagra (Madrid)		Metalurgica de Cubas	Pb
Barcelona		Derivados de Minerales y Metales	Pb
Medina del Campo (Valladolid)		Ferroaleaciones Espanolas	Pb
San Esteban de Gormaz (Soria)		Megorsa (Tudor)	Pb
Espinardo (Murcia)		Perdigones Azor	Pb
Barrio (Bilbao)		Fundiciones Ercoreco S.A.	Zn
Alcobendas (Madrid)		Fundiciones Triqueros	Zn
Cabrera de Mar (Barcelona)		Lainden S.A.	Zn
Asua (Bilbao)		Zincdes	Zn

UG = Underground mine; OP = Open pit mine; Sme = Smelter; Ref = Refinery
Source: Mining Magasine; Metaleurop, Annuaire statistique; Metal Bulletin; ILZSG.

Table 3.13 **Sweden: mining and processing facilities**

Name (location)	Type of operation	Company	Commodity	Production 1991 (tonnes) (metal content)	Capacity 1991 (tonnes)
Mines					
Aitik (Gällivare)	OP	Boliden Mineral AB	Ag	34	
			Cu	47 100	
Laisvall (Laisvall)	UG	Boliden Mineral AB	Ag	12	
			Pb	51 600	
			Zn	12 100	
Boliden	UG	Boliden Mineral AB	Ag	60	
			Cu	6 200	
			Pb	2 300	
			Zn	39 400	
Enåsen (Ljusdal)	UG	Boliden Mineral AB	Ag	2	
			Cu	1 400	
Garpenberg (Norra, Odal)	UG	Boliden Mineral AB	Ag	108	
			Cu	900	
			Pb	14 900	
			Zn	28 800	
Zinkgruvan (Åmmeberg)	UG	Vieille-Montagne Sverige	Ag	32	
			Pb	13 300	
			Zn	65 600	
Falun	OP/ UG	STORA	Ag	8	
			Cu	300	
			Pb	2 800	
			Zn	9 500	
Viscaria (Kiruna)	UG/ OP	Viscaria AB (Outokumpu Oy)	Ag	4	
			Cu	24 900	
Björkdal (Skellefbeå)	OP	Terra Mining AB	Au	2	
Smelters/Refineries					
Rönnskär (Skelleftehamn)	Sme-ref	Boliden Mineral AB	Ag	301	500
			Au	8.5	15
			Cu	96 650	120 000
			Zn	49 160	
			Pb	100 000	
Sundsvall	Sme	Gränges Aluminium Metall	Al	97 000	100 000
Secondary plants					
Ålmhult		AB Gotthard Aluminium	Al	14 800	16 000
Finspång		Finspång Aluminium (Gränges)	Al	52 000	90 000
Mölndal		Olof Manner AB	Al	1 400	2 000
Landskrona		Boliden Bergsöe AB	Cu (alloys)	5 950	
			Pb (alloys)	38 600	
			Sn (alloys)	1 264	

UG = Underground mine; OP = Open pit mine; Sme = Smelter; Ref = Refinery
Source: Ministry of Industry and Commerce.

Table 3.14 **United Kingdom: mining and processing facilities**

Name (location)	Company	Commodity	Capacity (tonnes metal)		Output (tonnes) (1989)
Mines					
Wheal Jane (Cornwall)*	Carnon Cons.	Ag			
		Cu			507
		Sn			895
		Zn			5 770
South Crofty (Cornwall)	Carnon Cons.	Sn			2 453
Parys Mt. (Anglesey)	Anglesey Mining	Ag			
		Cu	2 500	(planned)	
		Pb	5 700	(planned)	
		Zn	13 000	(planned)	
Smelters/Refineries					
Kinlochleven (Argyll, Scotland)	British Alcan Aluminium	Al	11 000		10 783
Lochaber (Invernesshire, Scotland)	British Alcan Aluminium	Al	38 000		36 806
Lynemouth (Ashington, Northumberland)	British Alcan Aluminium	Al	125 000		121 835
Holyhead (Gwynedd, Wales)	Anglesey Aluminium Metal	Al	125 000		123 100
Prescott (Merseyside)	BICC Metals	Cu	65 000		
North Ferriby (Humberside)**	Capper Pass & Son	Cu	4 000		
		Pb	8 000		
		Sn	15 000		10 768
Walsall		Cu	80 000		
Clydach (Wales)	INCO Europe	Ni	40 000		26 100
Northfleet (Kent)	Britannia Refined Metals	Ag			
		Pb	150 000		156 494
Avonmouth (Bristol)	Commonwealth Smelting Ltd	Pb	40 000		34 523
	(Pasminco Europe)	Zn	100 000		79 773

* Closure announced in March 1991

** Closed

Sources: UK Department of Trade and Industry; Mining Bulletin.

Table 3.15 **United States: mining and processing facilities**

Name (location)	Type of operation	Company	Commoditiy	Output ('000t)
Mines				
Bagdad (Arizona)	OP	Cyprus Bagdad Copper Co.	Cu	116
			Mo, Ag	
Bingham Canyon (Utah)	OP	Kennecott	Cu	255
			Ag, Au, Mo	
Casa Grande (Arizona)	UG	Cyprus Casa Grande Corp.	Cu	10
Chino (New Mexico)	OP	Phelps Dodge/Mitsubishi	Cu	150
			Ag, Mo	
Continental (Montana)	OP	Montana Resources Corp.	Cu, Ag, Mo	50
Inspiration (Arizona)	OP	Cyprus Miami Mining Corp.	Cu	65
Mission [Eisenhower/ Mission/Pima/ San Xavier](Arizona)	OP	Asarco Inc.	Cu, Ag	97
Morenci (Arizona)	OP	Phelps Dodge/Sumitomo	Cu	345
			Ag, Au, Mo	
Pinto Valley (Arizona)	OP	Magma Copper Co.	Cu	92
			Ag, Au, Mo	
Ray (Arizona)	OP	Asarco Inc.	Cu	125
			Ag	
San Manuel (Arizona)	OP-UG	Magma Copper Co.	Cu	130
			Ag, Au, Mo	
Sierrita/Twin Buttes (Arizona)	OP	Cyprus Minerals Co.	Cu, Mo	137
Silver Bell (Arizona)	OP	Asarco Inc.	Cu, Ag	5
Tyrone (New Mexico)	OP	Phelps Dodge	Cu, Ag, Au	110
White Pine (Michigan)	UG	Copper Range (Metall Mining)	Cu, Ag	60
Nickel Mountain Mine (Oregon)	OP	Several Parties	Ni	
Doe Run [Buick/Brushy Creek/Viburnum] (Missouri)	UG	The Doe Run Co.	Pb	310
			Ag	5
			Cu	9
			Zn	56
Leadville (Colorado)	UG	Asarco Incorporated	Pb	8
			Ag, Au	
			Zn	15
Lucky Friday Mine (Idaho)	UG	Hecla Mining Co.	Pb	29
			Zn	4
Rosiclare Works (Illinois)	UG	Ozark-Mahoning Co.	Pb	0.4
			Zn	3
Magmont (Missouri)	UG	Cominco American Inc.	Pb	80
			Ag, Cu	
			Zn	18
Sweetwater (Missouri)	UG	Asarco Inc.	Pb	91
			Ag	
			Zn	6
West Fork (Missouri)	UG	Asarco Inc.	Pb	65
			Ag	
			Zn	9
Balmat/Pierrepont (New York)	UG	Zinc Corp. of America	Zn	42
			Pb	0.5
Elmwood/Gordonsville (Tennessee)	UG	Jersey Miniere Zinc Co.	Pb	
			Cu	50

Table 3.15 (cont'd) **United States: mining and processing facilities**

Name (location)	Type of operation	Company	Commoditiy	Output ('000t)
Mines (cont'd)				
Greens Creek (Alaska)	UG	Greens Creek Mining (Kennecott)	Zn	30
			Ag, Au	
			Pb	10
Montana Tunnels (Montana)	OP	Pegasus Gold Corp.	Zn	22
			Ag, Au	
			Pb	5
Red Dog (Alaska)	OP	Cominco Alaska Inc.	Zn	325
			Ag	
			Pb	65
New Market/Young/ Immel/Coy (Tennessee)	UG	Asarco Inc.	Zn	44
Sunnyside Mine (Colorado)	UG	Washington Mining Co.	Pb	3
			Zn	5
Ward (Nevada) (Temp. Closed 1991)	UG	Alta Gold Co.	Zn	13
			Ag, Cu	
			Pb	1
Jefferson City (Tennessee)	UG	Union Zinc Inc.	Zn	16
Smelters/Refineries				Capacity ('000t)
St. Croix (Virgin Islands)	Ref	Virgin Islands Alumina	Al	635
Point Comfort (Texas)	Ref	Alcoa	Al	1 735
Corpus Christi (Texas)	Ref	Reynolds Metals Co.	Al	1 700
Gramercy (Louisiana)	Ref	Kaiser Aluminum & Chemical Corp.	Al	1 000
Burnside (Louisiana)	Ref	Ormet Corp.	Al	545
Alcoa (Tennessee)	Sme	Alcoa	Al	200
Badin (North Carolina)	Sme	Alcoa	Al	115
Warrick (Indiana)	Sme	Alcoa	Al	270
Massena (New York)	Sme	Alcoa	Al	127
Rockdale (Texas)	Sme	Alcoa	Al	310
Wenatchee (Washington)	Sme	Alcoa	Al	205
Columbia Falls (Montana)	Sme	Columbia Falls Aluminum Co.	Al	163
Sebree (Kentucky)	Sme	Alcan Aluminum Corp.	Al	163
Longview (Washington)	Sme	Reynolds Metals Co.	Al	204
St. Lawrence (New York)	Sme	Reynolds Metals Co.	Al	123
Troutdale (Oregon)	Sme	Reynolds Metals Co.	Al	121
Mead-Spokane (Washington)	Sme	Kaiser Aluminum & Chemical Corp.	Al	200
Tacoma (Washington)	Sme	Kaiser Aluminum & Chemical Corp.	Al	73
Ferndale (Washington)	Sme	Intalco (Alumax)	Al	254
Frederick (Maryland)	Sme	Eastalco (Alumax)	Al	160
Hannibal (Ohio)	Sme	Ormet Corp.	Al	245
Hawesville (Kentucky)	Sme	N.S.A.	Al	172
New Madrid (Missouri)	Sme	Noranda Aluminum Inc.	Al	204
The Dalles (Oregon)	Sme	Northwest Aluminum Corp.	Al	82
Goldendale (Washington)	Sme	Columbia Aluminium Corp.	Al	165
Mt. Holly (So Carolina)	Sme	Alumax Inc.	Al	181
Vancouver (Washington)	Sme	Vanalco Inc.	Al	110
Ravenswood (West Virginia)	Sme	Ravenswood Aluminum Corp.	Al	166

Table 3.15 (cont'd) **United States: mining and processing facilities**

Name (location)	Type of operation	Company	Commoditiy	Capacity ('000t)
Smelters/Refineries (cont'd)				
El Paso (Texas)	Sme	Asarco Inc.	Cu	104
Hayden (Arizona)	Sme	Asarco Inc	Cu	170
Amarillo (Texas)	Ref	Asarco Inc.	Cu	414
Ray (Arizona)	Ref	Asarco Inc.	Cu	45
Saint Louis (Missouri)	Ref	Cerro Copper Products Co.	Cu	80
White Pine (Michigan)	Sme	Copper Range Co.	Cu	60
	Ref			54
Bagdad (Arizona)	Ref	Cyprus Copper Co.	Cu	12
Lakeshore (Arizona)	Sme	Cyprus Casa Grande Corp.	Cu	45
	Ref			40
Globe (Arizona)	Sme	Cyprus Miami Co.	Cu	180
	Ref			110
Sierrita (Arizona)	Ref	Cyprus Sierrita Corp.	Cu	5.5
Twin Buttes (Arizona)	Ref	Cyprus Twin Buttes Corp.	Cu	20
Garfield (Utah)	Sme	Kennecott	Cu	210
	Ref			215
Baltimore (Maryland)	Ref	Cox Creek Refining Co.	Cu	100
San Manuel (Arizona)	Sme	Magma Copper Co.	Cu	272
	Ref			317
Miami (Arizona)	Ref	Pinto Valley Copper (Magma)	Cu	10
Pinto Valley (Arizona)	Ref	Pinto Valley Copper (Magma)	Cu	10
Hidalgo (New Mexico)	Sme	Phelps Dodge Corp.	Cu	180
Hurley (New Mexico)	Sme	Phelps Dodge Corp.	Cu	170
Chino (New Mexico)	Ref	Chino (Phelps Dodge)	Cu	45
Tyrone (New Mexico)	Ref	Burro Chief (Phelps Dodge)	Cu	50
Morenci (Arizona)	Ref	Phelps Dodge Corp.	Cu	100
El Paso (Texas)	Ref	Phelps Dodge Refining Corp.	Cu	440
Three Rivers (Michigan)	Ref	Essex Group	Cu	34
Sauget (Illinois)	Sme-2nd	Cerro Copper Products	Cu	70
	Ref		Cu	135
Alton (Illinois)	Sme-2nd	Concorde Trading Corp	Cu	135
Warrenton (Missouri)	Ref-2nd	Warrenton Refining Co. (Cyprus)	Cu	32
Philadephia (Pennsylvania)	Sme-2nd	Franklin Smelting & Refining Co.	Cu	16
Gaston (South Carolina)	Sme-2nd	Gaston Recycling	Cu	110
	Ref			120
Reading (Pennsylvania)	Ref-2nd	Reading Metals	Cu	70
Carrollton (Georgia)	Sme	Southwire Co. Inc.	Cu	100
	Ref	Southwire Co. Inc.	Cu	115
Riddle (Oregon)	Sme	Glenbrook Nickel Corp.	Ni	
East Helena (Montana)	Sme	Asarco Inc.	Pb	82
El Paso (Texas)	Sme	Asarco Inc.	Pb	
Omaha (Nebraska)	Ref	Asarco Inc.	Pb	164
Glover (Missouri)	Sme-Ref	Asarco Inc.	Pb	112
Herculaneum (Missouri)	Sme-Ref	The Doe Run Co.	Pb	204
Boss (Missouri)	Ref	The Doe Run Co.	Pb	127

Table 3.15 (cont'd) **United States: mining and processing facilities**

Name (location)	Type of operation	Company	Commoditiy	Capacity ('000t)
Smelters/Refineries (cont'd)				
Corpus Christi (Texas)	Ref	Asarco Inc.	Zn	
Sauget (Illinois)	Ref	Big River Zinc Corp.	Zn	82
Clarksville (Tennessee)	Ref	Jersey Miniere Zinc Co.	Zn	98
Monaca (Pennsylvania)	Ref	Zinc Corp. of America	Zn	146
Bartlesville (Oklahoma)	Ref	Zinc Corp. of America	Zn	51

UG = Underground mine; OP = Open pit mine; Sme = Smelter; Ref = Refinery; 2nd = Secondary plant
Source: US Bureau of Mines; Mining Magazine; International Lead and Zinc Study Group.

Table 4.1 **Australia: Post-Tokyo Round average tariffs on non-ferrous metals**

		Aluminium	Copper	Lead	Nickel	Tin	Zinc
Ores and concentrates[1]	S	1.5	1.5	1.5	1.5	1.5	1.5
	W	0.7	0.7	0.7	0.7	0.7	0.7
Unwrought	S	2.0	2.0	2.0	2.0	2.0	0.0
	W	2.0	2.0	2.0	2.0	2.0	0.0
Semi-manufactures	S	5.4	4.3	1.6	4.3	1.5	3.5
	W	10.9	8.1	1.9	2.0	2.0	2.3
Metal manufactures[1]	S	7.1	7.1	7.1	7.1	7.1	7.1
	W	13.6	13.6	13.6	13.6	13.6	13.6

1) Figures for both Ores and concentrates and for Metal manufactures are average rates for all non-ferrous metals
S: Simple average
W: Weighted average tariffs on basis of 1985-86 imports
Source: GATT, Trade Policy Review, Australia, March 1990.

Table 4.2 **EEC: Post-Tokyo Round average tariffs on non-ferrous metals**

		Aluminium	Copper	Lead	Nickel	Tin	Zinc
Ores and concentrates		0.0	0.0	0.0	0.0	0.0	0.0
Unwrought	S	2.8	0.0	1.2	0.0	0.0	1.7
	W	5.2	0.0	1.3	0.0	0.0	3.4
Semi-manufactures	S	8.4	5.2	7.2	3.7	3.6	7.1
	W	9.8	5.9	7.8	3.9	3.3	7.7
Metal manufactures[1]	S	5.6	5.6	5.6	5.6	5.6	5.6
	W	5.8	5.8	5.8	5.8	5.8	5.8

1) Metal manufactures include all non-ferrous metals
S: Simple average
W: Weighted average, calculated using MFN imports from late 1970s.
Source: GATT, 1984-1988, Background Studies on Trade in Natural Resource Products

Table 4.3 **Japan: Post-Tokyo Round average tariffs on non-ferrous metals**

		Aluminium	Copper	Lead	Nickel	Tin	Zinc
Ores and concentrates		0.0	0.0	0.0	0.0	0.0	0.0
Unwrought	S	6.0	4.6	5.0	4.5	1.1	2.9
	W	8.0	5.3	4.4	1.4	0.0	4.0
Semi-manufactures	S	9.6	7.0	7.2	5.5	4.3	5.4
	W	10.7	6.4	6.3	5.6	4.5	5.4
Metal manufactures[1]	S	5.4	5.4	5.4	5.4	5.4	5.4
	W	5.2	5.2	5.2	5.2	5.2	5.2

1) Metal manufactures include all non-ferrous metals
S: Simple average
W: Weighted average, calculated using MFN imports from late 1970s.
Source: GATT, 1984-1988, Background Studies on Trade in Natural Resource Products

Table 4.4 **Norway: Post-Tokyo Round average tariffs on non-ferrous metals**

		Aluminium	Copper	Lead	Nickel	Tin	Zinc
Ores and concentrates		0.0	0.0	0.0	0.0	0.0	0.0
Unwrought	S	0.0	0.0	0.0	0.0	0.0	0.0
	A	0.0	0.0	0.0	0.0	0.0	0.0
Semi-manufactures	S	3.7	1.5	0.0	0.7	0.8	0.4
	A	0.2	0.0	0.0	0.1	0.0	0.0
Metal manufactures[1]	S	4.8	4.8	4.8	4.8	4.8	4.8
	A	0.5	0.5	0.5	0.5	0.5	0.5

1) Metal manufactures include all non-ferrous metals
S: Simple average
A: Applied basis
Source: GATT, Trade Policy Review Mechanism, Norway, August 1991

Table 4.5 **Sweden: Post-Tokyo Round average tariffs on non-ferrous metals**

		Aluminium	Copper	Lead	Nickel	Tin	Zinc
Ores and concentrates		0.0	0.0	0.0	0.0	0.0	0.0
Unwrought	S	0.0	0.0	0.0	0.0	0.0	0.0
	W	0.0	0.0	0.0	0.0	0.0	0.0
Semi-manufactures	S	2.4	2.5	0.0	1.0	0.3	0.0
	W	2.5	2.4	0.0	0.7	1.0	0.0
Metal manufactures[1]	S	3.8	3.8	3.8	3.8	3.8	3.8
	W	3.9	3.9	3.9	3.9	3.9	3.9

1) Metal manufactures include all non-ferrous metals
S: Simple average
W: Weighted average, calculated using MFN imports from late 1970s.
Source: GATT, 1984-1988, Background Studies on Trade in Natural Resource Products

Table 4.6 **United States: Post-Tokyo Round average tariffs on non-ferrous metals**

		Aluminium	Copper	Lead	Nickel	Tin	Zinc
Ores and concentrates[1]	S	0.0	0.8	1.9	0.0	0.0	1.5
	W	0.0	0.4	2.2	0.0	0.0	2.5
Unwrought	S	1.3	2.7	3.1	0.0	0.0	7.5
	W	0.1	0.9	3.4	0.0	0.0	1.5
Semi-manufactures	S	3.9	4.2	3.9	5.1	3.8	4.7
	W	3.3	1.6	2.4	1.2	3.3	2.1
Metal manufactures[2]	S	7.9	7.9	7.9	7.9	7.9	7.9
	W	4.5	4.5	4.5	4.5	4.5	4.5

1) *Ad valorem* incidence of specific rates based on 1981 (copper) or 1984 (lead and zinc) trade
2) Metal manufactures include all non-ferrous metals
S: Simple average
W: Weighted average, calculated using MFN imports from late 1970s
Source: GATT, 1984-1988, Background Studies on Trade in Natural Resource Products

Table 5.1 **Australia: tariff treatment according to stage of processing of six non-ferrous metals**

METAL/ Stage of processing	Tariff heading No.	MFN[a] (In %, unbound, except for *)
ALUMINIUM		
Ores and concentrates	26.06	0.0
Unwrought	76.01	0.0
— Waste and scrap	76.02	0.0
— Powders and flakes	76.03	10.0
Wrought		
— Semis	76.04-.07	10;15
— Tubes, fittings	76.08-.09	10;15
Chemicals	28.18-.42	0;10
Finished manufactures	76.10-.16	15.0
COPPER		
Ores and concentrates	26.03;.20-.21	0.0
Unwrought	74.01-.05	0.0
— Powders and flakes	74.06	0.0
Wrought	74.07-.12	10;15
Chemicals	28.25-.42	0;10
Finished manufactures	74.13-.19	10;15*
LEAD		
Ores and concentrates	26.07;.20-.21	0.0
Unwrought	78.01	0.0
— Waste and scrap	78.02	0.0
— Powders and flakes	78.04	0.0
Wrought	78.03-.05	0.0
Chemicals	28.24-.42	0.0
Finished manufactures	78.06	15.0
NICKEL		
Ores and concentrates	26.04	0.0
Unwrought	75.01-.02	0.0
— Waste and scrap	75.03	0.0
— Powders and flakes	75.04	0.0
Wrought	75.05-.07	0;15
Chemicals	28.25-.42	0.0
Finished manufactures	75.08	0.0

Table 5.1 (cont'd) **Australia: tariff treatment according to stage of processing of six non-ferrous metals**

METAL/ Stage of processing	Tariff heading No.	MFN[a] (In %, unbound, except for *)
TIN		
Ores and concentrates	26.09	0.0
Unwrought	80.01	0.0
— Waste and scrap	80.02	0.0
— Powders and flakes	80.05.20	0.0
Wrought	80.03-.05.10	0.0
Chemicals	28.25-.42	0.0
Finished manufactures	80.07	0.0
ZINC		
Ores and concentrates	26.08	0.0
Unwrought	79.01	0.0
— Waste and scrap	79.02	0.0
— Powders and flakes	79.03	0;10
Wrought	79.04-.06	0.0
Chemicals	28.17-.42	0;10
Finished manufactures	79.07	15.0

a) Rates of 10 per cent and 15 per cent will be progressively reduced to 5 per cent by 1 July 1996.

Table 5.2 **Canada: tariff treatment according to stage of processing of six non-ferrous metals**

METAL/ Stage of Processing	Tariff heading No.	MFN (applied rates in per cent unless otherwise indicated)
ALUMINIUM		
Ores and concentrates	26.06	0.0
Unwrought	76.01	0 - 10.3; 1.98¢/kg
— Waste and scrap	76.02	0.0
— Powders and flakes	76.03	9.2 - 10.3
Wrought		
— Semis	76.04-.07	0 - 12.2
— Tubes, fittings	76.08-.09	8.1 - 10.3
Chemicals	28.18-.42	0 - 12.5
Finished manufactures	76.10-.16	0 - 11.4
COPPER		
Ores and concentrates	26.03	0.0
Unwrought		
— Mattes	74.01	0.0
— Unrefined + refined	74.02-.03	0 - 10.3
— Waste & scrap	74.04	0 - 10.2
— Master alloys	74.05	10.3
— Powders and flakes	74.06	4 - 10.6
Wrought	74.07-.12	0 - 10.3
Chemicals	28.25-.42	0 - 12.5
Finished manufactures	74.13-.19	0 - 15.0
LEAD		
Ores and concentrates	26.07	0.0
Unwrought	78.01	0 - 10.2
— Waste and scrap	78.02	0.0
— Powders and flakes	78.04	4 - 10.2
Wrought	78.03-.05	4 - 10.2
Chemicals	28.24-.42	0 - 12.5
Finished manufactures	78.06	10.2
NICKEL		
Ores and concentrates	26.04	0.0
Unwrought	75.01-.02	0.0
— Waste and scrap	75.03	0.0
— Powders and flakes	75.04	0 - 10.2
Wrought	75.05-.07	4.2 - 10.2
Chemicals	28.25-.42	0 - 12.5
Finished manufactures	75.08	0 - 11.0

Table 5.2 (cont'd) **Canada: tariff treatment according to stage of processing of six non-ferrous metals**

METAL/ Stage of processing	Tariff heading No.	MFN (applied rates in per cent unless otherwise indicated)
TIN		
Ores and concentrates	26.09	0.0
Unwrought	80.01	0 - 10.2
— Waste and scrap	80.02	0.0
— Powders and flakes	80.05	0 - 10.2
Wrought	80.03	0 - 10.2
Chemicals	28.25-.42	0 - 12.5
Finished manufactures	80.07	10.2
ZINC		
Ores and concentrates	26.08	0.0
Unwrought	79.01	0 - 17.5
— Waste and scrap	79.02	0.0
— Powders and flakes	79.03	0 - 10.2
Wrought	79.04-.06	0 - 10.2
Chemicals	28.17-.42	0 - 12.5
Finished manufactures	79.07	0 - 10.2

Table 5.3 **EEC: tariff treatment according to stage of processing of six non-ferrous metals**

METAL/ Stage of Processing	Tariff heading No.	MFN (applied rates in per cent)
ALUMINIUM		
Ores and concentrates	26.06	0.0
Unwrought	76.01	6.0
— Waste and scrap	76.02	0;3.2
— Powders and flakes	76.03	5.3;6.3
Wrought		
— Semis	76.04-.07	7.5-10
— Tubes, fittings	76.08-.09	0-10
Chemicals	28.18-.42	3-12
Finished manufactures	76.10-.16	7.0
COPPER		
Ores and concentrates	26.03;.20-.21	0.0
Unwrought	74.01-.05	0.0
— Powders and flakes	74.06	1.4;6.2
Wrought	74.07-.12	5.0-6.0;4.5
Chemicals	28.25-.42	3.2-12
Finished manufactures	74.13-.19	0;4.9;6.5
LEAD		
Ores and concentrates	26.07;.20-.21	0.0
Unwrought	78.01	0;3.5
— Waste and scrap	78.02	0.0
— Powders and flakes	78.04	2.2
Wrought	78.03-.05	8;9.0
Chemicals	28.24-.42	3.2-12
Finished manufactures	78.06	6.0;8.0
— Batteries	85.07	5.6;6.2
NICKEL		
Ores and concentrates	26.04	0.0
Ferronickel	72.02	0.0
Unwrought	75.01-.02	0.0
— Waste and scrap	75.03	0.0
— Powders and flakes	75.04	0.5
Wrought	75.05-.07	3.8-5.3
Chemicals	28.25-.42	0-12
Finished manufactures	75.08	4.6

Table 5.3 (cont'd) **EEC: tariff treatment according to stage of processing of six non-ferrous metals**

METAL/ Stage of processing	Tariff heading No.	MFN (applied rates in per cent)
TIN		
Ores and concentrates	26.09	0.0
Unwrought	80.01	0.0
— Waste and scrap	80.02	0.0
— Powders and flakes	80.05.20	2.9
Wrought	80.03-.05	2.5-4.5
Chemicals	28.25-.42	3-12
Finished manufactures	80.07	5.3
ZINC		
Ores and concentrates	26.08	0.0
Unwrought	79.01	3.5
— Waste and scrap	79.02	0.0
— Powders and flakes	79.03	4.4
Wrought	79.04-.06	8.0
Chemicals	28.17-.42	3-12
Finished manufactures	79.07	7.0

Table 5.4 **Japan: tariff treatment according to stage of processing of six non-ferrous metals**

METAL/ Stage of processing	Tariff heading No.	MFN (applied rates in per cent unless otherwise indicated)
ALUMINIUM		
Ores and concentrates	26.06	0.0
Unwrought	76.01	1.0
— Waste and scrap	76.02	0.0
— Powders and flakes	76.03	4.6
Wrought	76.04-.09	0-10.2
Chemicals	28.16-.42	3-4.6
Finished manufactures	76.10-.16	4.1-6.3
COPPER		
Ores and concentrates	26.03	0.0
Unwrought	74.01-.06	0 ; 4.8-5.8 ; 15 ¥/kg[1]
— Mattes	74.01	0.0
— Waste and scrap	74.04	0.0
— Powders and flakes	74.06	5.8
Wrought	74.07-.12	4.6-6.6
Chemicals	28.25-.42	3.9-5.8
Finished manufactures	74.13-.19	3.9-6.8
LEAD		
Ores and concentrates	26.07	0.0
Unwrought	78.01	0 ;2.8-5.2 ; 8 ¥/kg;[1]
— Waste and scrap	78.02	2.6
— Powders and flakes	78.04	5.2-6.6
Wrought	78.03-.05	4.6-6.6
Chemicals	28.24-.42	3-5.6
Finished manufactures	78.06	4.6
NICKEL		
Ores and concentrates	26.04	0.0
Unwrought	75.01-.02	0 ; 5.8 ; 8.1 ;[2]
— Waste and scrap	75.03	0.0
— Powders and flakes	75.04	0;4.8;52 ¥/kg
Wrought	75.05-.07	0 ; 4.6 ;5.8
Chemicals	28.25-.42	3-5.8
Finished manufactures	75.08	4.6

Table 5.4 (cont'd) **Japan: tariff treatment according to stage of processing of six non-ferrous metals**

METAL/ Stage of processing	Tariff heading No.	MFN (applied rates in per cent unless otherwise indicated)
TIN		
Ores and concentrates	26.09	0.0
Unwrought	80.01	0;2.6
— Waste and scrap	80.02	0.0
— Powders and flakes	80.05.20	3.9
Wrought	80.03.-.05	3;3.9
Chemicals	28.25-.42	3.9
Finished manufactures	80.07	4.6
ZINC		
Ores and concentrates	26.08	0.0
Unwrought	79.01	0 ; 5.76-6.24 ; 8 ¥/kg[1]
— Waste and scrap	79.02	0.0
— Powders and flakes	79.03	4.6
Wrought	79.04-.06	3.8 ; 5.8
Chemicals	28.17-.42	3-4.6
Finished manufactures	79.07	3.9 ; 4.6

1) Duties on certain forms of the unwrought metal increase/decrease with decrease/increase in prices of refined metal.
2) Lesser of 11.7 per cent or 72.9 ¥/kg.

Table 5.5 **Norway: tariff treatment according to stage of processing of six non-ferrous metals**

METAL/ Stage of Processing	Tariff heading No.	MFN (applied rates in per cent unless otherwise indicated)
ALUMINIUM		
Ores and concentrates	26.06	0.0
Unwrought	76.01	0.0
— Waste and scrap	76.02	0.0
— Powders and flakes	76.03	0.0
Wrought		
— Semis	76.04-.07	0 - 5.8; NKr 1.00/kg
— Tubes, fittings	76.08-.09	5.8 - 6.2
Chemicals	28.18-.42	0; 7.7
Finished manufactures	76.10-.16	3.8 - 13; NKr 0.16-1.12/kg
COPPER		
Ores and concentrates	26.03	0.0
Unwrought		
— Mattes + cement	74.01	0.0
— Unrefined + refined	74.02-.03	0.0
— Waste & scrap	74.04	0.0
— Master alloys	74.05	0.0
— Powders and flakes	74.06	0.0
Wrought	74.07-.12	0 - 5.1; NKr 0.06/kg
Chemicals	28.25-.48	0; 7.7
Finished manufactures	74.13-.19	0 - 5.1; NKr 0.10/kg
LEAD		
Ores and concentrates	26.07	0.0
Unwrought	78.01	0.0
— Waste and scrap	78.02	0.0
— Powders and flakes	78.04.20	0.0
Wrought	78.03-.05	0.0
Chemicals	28.24-.42	0; 7.7
Finished manufactures	78.06	3.2
NICKEL		
Ores and concentrates	26.04	0.0
Unwrought	75.01-.02	0.0
— Waste and scrap	75.03	0.0
— Powders and flakes	75.04	0.0
Wrought	75.05-.07	0; 2.2
Chemicals	28.25-.42	0.0
Finished manufactures	75.08	2.5

Table 5.5 (cont'd) **Norway: tariff treatment according to stage of processing of six non-ferrous metals**

METAL/ Stage of processing	Tariff heading No.	MFN (applied rates in per cent unless otherwise indicated)
TIN		
Ores and concentrates	26.09	0.0
Unwrought	80.01	0.0
—Waste and scrap	80.02	0.0
—Powders and flakes	80.05.20	0.0
Wrought	80.03-.05.10;.06	0; 3.8
Chemicals	28.25-.42	0; 7.7
Finished manufactures	80.07	3.2
ZINC		
Ores and concentrates	26.08	0.0
Unwrought	79.01	0.0
— Waste and scrap	79.02	0.0
— Powders and flakes	79.03	0.0
Wrought	79.04-.06	0;2.2
Chemicals	28.17-.42	0 ; NKr 0.14/kg
Finished manufactures	79.07	3.8

Table 5.6 **Sweden: tariff treatment according to stage of processing of six non-ferrous metals**

METAL/ Stage of processing	Tariff heading No.	MFN (applied rates in per cent)
ALUMINIUM		
Ores and concentrates	26.06	0.0
Unwrought	76.01	0.0
— Waste and scrap	76.02	0.0
— Powders and flakes	76.03	0; 3.8
Wrought		
— Semis	76.04-.09	2.5
— Tubes, fittings	76.08-.09	2.5; 3.2
Chemicals	28.16-.42	0 - 5.8
Finished manufactures	76.10-.16	2.5-3.8
COPPER		
Ores and concentrates	26.03;.20-.21	0.0
Unwrought	74.01-.06	0.0
— Powders and flakes	74.06	0; 3.8
Wrought	74.07-.12	1.4; 2.5; 3.2
Chemicals	28.25-.42	0; 5.8
Finished manufactures	74.13-.19	2.5; 3.2
LEAD		
Ores and concentrates	26.07	0.0
Unwrought	78.01	0.0
— Waste and scrap	78.02	0.0
— Powders and flakes	78.04	0.0
Wrought	78.03-.05	0.0
Chemicals	28.24-.42	0; 3.2; 5.8
Finished manufactures	78.06	0; 3.2; 3.8
NICKEL		
Ores and concentrates	26.04	0.0
Unwrought	75.01-.02	0.0
— Waste and scrap	75.03	0.0
— Powders and flakes	75.04	0.0
Wrought	75.05-.07	1.4
Chemicals	28.25-.42	0; 5.8
Finished manufactures	75.08	0; 3.2

Table 5.6 (cont'd) **Sweden: tariff treatment according to stage of processing of six non-ferrous metals**

METAL/ Stage of processing	Tariff heading No.	MFN (applied rates in per cent)
TIN		
Ores and concentrates	26.09	0.0
Unwrought	80.01	0.0
—Waste and scrap	80.02	0.0
—Powders and flakes	80.05.20	0.0
Wrought	80.03,-.05.10 ;.06	0; 1.4
Chemicals	28.25-.42	0; 3.2; 5.8
Finished manufactures	80.07	3.2; 3.8
ZINC		
Ores and concentrates	26.08	0.0
Unwrought	80.01	0.0
— Waste and scrap	79.02	0.0
— Powders and flakes	79.03	0.0
Wrought	79.04-.06	0; 2.2
Chemicals	28.17-.42	0; 5.8
Finished manufactures	79.07	3.8

Table 5.7 **United States: tariff treatment according to stage of processing of six non-ferrous metals**

METAL/ Stage of processing	Tariff heading No.	MFN (applied rates; in per cent unless otherwise indicated)
ALUMINIUM		
Ores and concentrates	26.06	0.0
Unwrought	76.01	0 - 2.6
— Waste and scrap	76.02	0.0
— Powders and flakes	76.03	3.9; 5.7
Wrought		
— Semis	76.04-.07	0-6.5
— Tubes, fittings	76.08-.09	5.7
Chemicals	28.18-.42	0 - 7
Finished manufactures	76.10-.16	0 - 6.3
COPPER		
Ores and concentrates	26.03	1.7¢/kg on Pb content
Unwrought		
— Mattes	74.01	0.7¢/kg on both Cu & on Pb content
Cement copper		1.7% on value of Cu content
— Unrefined + refined	74.02-.03	1.0
— Waste & scrap	74.04	0.0
— Master alloys	74.05	2.6; 6
— Powders and flakes	74.06	3; 5.4
Wrought	74.07-.12	1 - 11.2
Chemicals	28.25-.48	0 - 16
Finished manufactures	74.13-.19	0 - 10
LEAD		
Ores and concentrates	26.07	1.7¢/kg on Pb content + 0.7¢/kg on Zn content
Unwrought	78.01	3.0 on value of Pb content
— Waste and scrap	78.02	2.3 on value of Pb content
— Powders and flakes	78.04	11.25
Wrought	78.03-.05	1.2 - 3.4
Chemicals	28.24-.42	0 - 7
Finished manufactures	78.06	3.9
NICKEL		
Ores and concentrates	26.04	0.0
Unwrought	75.01-.02	0.0
— Waste and scrap	75.03	0.0
— Powders and flakes	75.04	0.0
Wrought	75.05-.07	3.3 - 5.5
Chemicals	28.25-.42	0 - 3.7
Finished manufactures	75.08	4.7; 5.5

Table 5.7 (cont'd) **United States: tariff treatment according to stage of processing of six non-ferrous metals**

METAL/ Stage of processing	Tariff heading No.	MFN (applied rates in per cent unless otherwise indicated)
TIN		
Ores and concentrates	26.09	0.0
Unwrought	80.01	0.0
— Waste and scrap	80.02	0.0
— Powders and flakes	80.05.20	4.2
Wrought	80.03-.05.10;.06	2.4; 4.2
Chemicals	28.25-.42	0 - 7
Finished manufactures	80.07	3.1; 4.2
ZINC		
Ores and concentrates	26.08	1.7¢/kg on Pb content + 0.7¢/kg on Zn content
Unwrought	79.01	1.5; 19
— Waste and scrap	79.02	0.0
— Powders and flakes	79.03	0.7¢/kg; 9.5%
Wrought	79.04-.06	3.8; 4.2
Chemicals	28.17-.42	0 - 7
Finished manufactures	79.07	3.4; 5.7

Table 5.8 **Canada/United States: selected list of mineral and metal tariffs under the Canada-United States Free Trade Agreement**

Short Description	CANADA	USA	PHASE-OUT
	(Tariffs as of 31 December 1988) (% unless otherwise noted)		1 January of year indicated
Metallic Ores and Concentrates (H.S. Chapter 26)			
Most ores & concentrates	0	0	-
Lead	0	1.7¢/kg	1998
Zinc	0	0.7¢/kg	1998
Intermediate Metallic Compounds (H.S. Chapter 28)			
Lead oxides	10.5/12.5	2.4/15	1993
Zinc oxides	10.5	0	1998
Ferroalloys (H.S. Chapter 72)			
Ferronickel	10.2	0	1989
Smelted and Refined Metals (H.S. Chapter 28,74-81)			
Aluminium			
Metal	0	2.6	1989
Waste and scrap	0	0	-
Powders and flakes	9.2/10.3	3.9/5.7	1993
Copper			
Refined	0/10.3	1	1993
Alloys	0/10.2	1	1993
Waste and scrap	0/10.2	0	1993
Master alloys	10.3	2.6 & 6	1993
Powders and flakes	4 & 10.6	3/5.4	1998
Nickel			
Refined and alloys	0	0	-
Powders and flakes	0 & 10.2	0	1993
Lead			
Refined and other	0/10.2	3 & 3.5	1998
Waste and scrap	0	2.3	1989
Powders and flakes	4 & 10.2	11.25	1998
Zinc			
Refined	0	1.5	1998
Alloys	0 & 17.5	19	1998
Waste and scrap	0	2.1	1989
Powders and flakes	4 & 10.2	0.7¢/kg & 9.5	1998
Dust	0	0.7¢/kg	1998
Tin			
Refined	0	0	-

Source: Energy, Mines and Resources Canada

Table 6.1 **EC: production and consumption of six major non-ferrous metals, average 1990/1991** (per cent share of market economies)

	ALUMINIUM				
	Production				Consumption
	Bauxite	Alumina	Aluminium	Semis[a]	Refined
Belgium/Luxembourg				2.2	1.8
Denmark				0.1	0.1
France	0.3	1.3	2.3	5.7	4.0
Western Germany		2.8	5.2	10.5	7.3
Greece	2.1	1.5	2.1	0.8	0.6
Ireland		2.3			
Italy		1.9	1.8	6.4	3.7
Netherlands			1.7	1.0	0.6
Portugal					0.3
Spain		2.4	1.9	2.2	1.6
United Kingdom		0.3	1.8	2.5	2.4
EC total	2.4	12.5	16.8	31.4	22.4

	COPPER				
	Production				Consumption
	Mine	Smelter[b]	Refined	Semis[c]	Refined
Belgium/Luxembourg		1.0	2.9	4.5[d]	3.5
Denmark					
France		0.1	0.5	6.7	4.4
Western Germany		2.6	4.7	12.4	9.4
Greece				0.6	0.5
Ireland					
Italy			0.8	7.8	4.4
Netherlands					0.2
Portugal	1.7			0.4	0.2
Spain	0.1	1.6	1.7	1.9	1.4
United Kingdom			0.9	3.8	2.7
EC total	1.9	5.2	11.5	38.1	26.7

	LEAD			ZINC		
	Production		Consumption	Production		Consumption
	Mine	Refined	Refined	Mine	Smelter[c]	Refined
Belgium/Luxembourg		1.7	1.2		4.1	2.7
Denmark			0.1			0.2
France	0.1	4.8	4.6	0.3	3.9	4.1
Western Germany	0.2	6.7	7.8	0.8	4.9	7.7
Greece	0.9		0.3	0.4		0.3
Ireland	1.1	0.2	0.4	2.4		
Italy	0.5	3.4	4.7	0.5	3.5	4.0
Netherlands		0.6	1.1		2.9	1.1
Portugal		0.1	0.4		0.1	0.2
Spain	1.7	2.1	2.3	3.5	3.6	1.7
United Kingdom		5.7	5.1	0.1	1.4	2.7
EC total	4.4	25.4	27.9	7.9	24.3	24.7

Table 6.1 (cont'd) **EC: production and consumption of six major non-ferrous metals, average 1990/1991** (per cent share of market economies)

	NICKEL			TIN		
	Production		Consumption	Production		Consumption
	Mines	Refined	Refined	Mines	Refined	Refined
Belgium/Luxembourg			2.3		0.1	1.0
Denmark			0.1			
France	10.1[f]	4.4[g]	4.6			3.6
Western Germany		0.1	9.6	0.5	0.9	9.1
Greece	1.9	1.7			0.1	0.4
Ireland						
Italy			3.3			2.5
Netherlands			0.2		2.5	2.9
Portugal				1.1		0.3
Spain			2.1		0.7	1.5
United Kingdom		3.0	3.5	1.4	3.9	4.5
EC total	12.0	9.1	25.7	3.1	8.2	25.7

a) 1990
b) Blister only
c) Including copper alloy semis
d) Including Netherlands
e) Slab zinc
f) New Caledonia only
g) Including New Caledonia

Source: Metalli Non Ferrosi, Nuova SAMIM; World Metal Statistics.

Table 6.2 **EC: employment in ore mining and non-ferrous metals manufacturing** ('000)

	Year	Average number of employees		Total employment	Share of (A) + (B) in total employment (per cent)
		Ore mining (A)	Non-ferrous metals manuf. (B)		
Belgium	1985	0	64*	3 561	1.80
Denmark	1989	0	1.7	2 645	0.06
West Germany	1989	1	103	27 742	0.37
France	1989	7	49	21 761	0.26
Greece	1986	4	6	3 600	0.28
Ireland	1986	1	0.3	1 081	0.12
Italy	1987	5	40	20 986	0.21
Luxembourg	1988	0	1.4	175	0.80
Netherlands	1989	0	33*	6 155	0.54
Portugal	1987	2	3	4 403	0.11
Spain	1988	5	22	11 773	0.23
United Kingdom	1988	0	59	25 914	0.23
EC total		25	382.4	129 796	0.31**

(A) ISIC 230
(B) ISIC 372
* Includes ISIC 371, production of iron and steel
** Both for the weighted average using totals for the 12 and the unweighted average after eliminating the highest and lowest shares.

Source: Industrial Statistics Yearbook 1989, UN; Yearbook of Labour Statistics 1992, ILO.

Table 6.3 **EC: exports and imports of non-ferrous metals and semis (excluding ores), 1990** (million US$)

	Copper	Nickel	Aluminium	Lead	Zinc	Tin
Exports						
Belgium/Luxembourg	1 829	13	1 272	75	373	24
Denmark	28	0	150	1	1	0
France	1 287	144	1 689	58	231	3
Western Germany	2 590	320	3 387	97	244	31
Greece	73	0	279	0	1	0
Ireland	6	11	26	17	0	0
Italy	733	26	767	13	94	6
Netherlands	272	9	1 473	24	285	26
Portugal	26	0	33	0	1	0
Spain	214	3	381	17	195	3
United Kingdom	630	365	1 046	133	49	55
Total						
A (including intra-EC)	7 689	891	10 505	438	1 474	149
B (excluding intra-EC)	2 199	374	2 815	89	351	41
B/A	29%	42%	27%	20%	24%	28%
Imports						
Belgium/Luxembourg	1 491	93	1 313	43	110	19
Denmark	156	6	267	12	39	2
France	2 312	413	2 532	61	228	62
Western Germany	2 981	660	3 768	135	493	152
Greece	202	2	86	17	40	6
Ireland	103	19	170	8	4	2
Italy	2 062	202	1 784	112	171	54
Netherlands	502	34	1 136	48	64	20
Portugal	165	4	191	22	35	5
Spain	555	123	377	14	20	23
United Kingdom	1 472	297	1 777	225	215	46
Total						
A (including intra-EC)	12 003	1 852	13 402	697	1 420	390
B (excluding intra-EC)	6 394	1 290	5 291	342	380	247
B/A	53%	70%	39%	49%	27%	63%

Source: OECD trade data: SITC sections 682-687.

Table 6.4. **EC: Selected EC World Mining Interests**

Country / Mineral	Western Germany	France	Italy	Netherlands	United Kingdom	Belgium
Bauxite	Halco (Guinea) Frialco (Guinea)	Halco (Guinea) Frialco (Guinea)	Halco (Guinea) Madras (India)	Mineração Rio do Norte (BRA) NV Billiton (SUR) Halco (Guinea) Worsley (AUS)	Comalco (AUS)	
Copper	OK Tedi (PNG) Teck Corp (CAN) Afton Mining (CAN) Highmont (CAN) MIM (AUS) Copper Range (USA)			Cuajone (Peru)	Tsumeb (Namibia) Mines Selbaie(CAN) Kennecott (USA) Magma Copper (USA) Pinto Valley (USA) Renison GC (AUS) Lornex Mining (CAN) Bougainville Copper (PNG) Palabora (ZAF) Olympic Dam (AUS) O'okiep Copper (ZAF) So.Peru Copper(Peru)	
Lead	Kemco (Thailand) MIM (AUS)			Pering Mine (ZAF) Cadje-but (AUS)	Black Mount. (ZAF) AM & S (AUS)	Djebel Aouam (Morocco)
Nickel				Cerro Matoso (Colombia)	Botswana RST (BWA) Agnew (AUS)	
Tin	PT Preussag (Indonesia)	Somika (Zaire) Serem Malaysia (MYS) Mayo Darle (Cameroon)			MMC (MYS) Renison GC (AUS) East Kemptville Tin Corp (CAN) Cesbra (BRA)	
Zinc	Kemco (Thailand) Teck Corp (CAN) MIM (AUS) Cominco (CAN)	Buenaventura (Peru)		Pering Mine (ZAF)	Black Mount. (ZAF) AM & S (AUS)	Union Mines (USA) Vieille Montagne (SWE) Asturienne (SWE)

Notes: MYS - Malaysia; ZAF - Republic of South Africa; SUR - Suriname; PNG - Papua New Guinea.
Source: Eurométaux, Conditions of Profitability of the EC Mining Industry; D. HUMPHREYS (1988), *Resources Policy,* March.

Table 6.5 **EC: SYSMIN under Lomé II, III and IV**

Country (Commodity)	Year	Commitment (ECU million)	Disbursement (ECU million)	Purpose
LOMÉ II (1980-1985) ECU 282 million available				
Zaire (copper, cobalt)	1981	40	40.0	Rehabilitation of mine equipment (Gécamines)
Zambia (copper, cobalt)	1981	55	55.0	Rehabilitation of mine equipment (ZCCM)
Rwanda (tin)	1984	2.8	0.5	Training
Guyana (bauxite)	1984	34.5	25.8	Purchase of parts & rehabilitation of mine equipment
Zambia II (copper, cobalt)	1985	28	27.2	Rehabilitation of mine equipment (ZCCM)
Zaire II (copper, cobalt)	1985	41	33.0	Rehabilitation of mine equipment (Gécamines/SNCZ)
Jamaica (bauxite/alumina)	1987			Rehabilitation
Liberia (iron ore)	1986	[a]	3.6	Rehabilitation of the Bond Mining Company
Sierra Leone (iron ore)		[b]		
TOTAL		201	185	
Lomé III (1986-1990) ECU 415 million available				
Mauritania (iron ore)	1987	18	13.4	Rehabilitation of railway (SNIM)
Niger (gold, coal)	1986	12.5	5.3	Exploration
Guinea (alumina)	1987	35	14.0	Modernisation of refinery (FRIGUIA)
Togo (phosphates)	1988	15.7	-	R&D and productivity improvements (OTP)
Botswana (nickel, copper)	1989	21.7	-	Modernisation of mine (BCL)
Zambia III (copper, cobalt)		[a]		
Senegal (phosphates)		15.0	-	
Papua New Guinea (copper)		18.0	-	
TOTAL		136	33	
Lomé IV (1991-1995) ECU 480 million available				
Namibia [c]	1991			Aid programme for the mineral sector
Zambia [c]	1992			Diversification
Dominican Republic [c]	1992			Exploration

[a] Cancelled
[b] Request rejected
[c] Under consideration

MAIN SALES OUTLETS OF OECD PUBLICATIONS
PRINCIPAUX POINTS DE VENTE DES PUBLICATIONS DE L'OCDE

ARGENTINA – ARGENTINE
Carlos Hirsch S.R.L.
Galería Güemes, Florida 165, 4° Piso
1333 Buenos Aires Tel. (1) 331.1787 y 331.2391
Telefax: (1) 331.1787

AUSTRALIA – AUSTRALIE
D.A. Information Services
648 Whitehorse Road, P.O.B 163
Mitcham, Victoria 3132 Tel. (03) 873.4411
Telefax: (03) 873.5679

AUSTRIA – AUTRICHE
Gerold & Co.
Graben 31
Wien I Tel. (0222) 533.50.14

BELGIUM – BELGIQUE
Jean De Lannoy
Avenue du Roi 202
B-1060 Bruxelles Tel. (02) 538.51.69/538.08.41
Telefax: (02) 538.08.41

CANADA
Renouf Publishing Company Ltd.
1294 Algoma Road
Ottawa, ON K1B 3W8 Tel. (613) 741.4333
Telefax: (613) 741.5439
Stores:
61 Sparks Street
Ottawa, ON K1P 5R1 Tel. (613) 238.8985
211 Yonge Street
Toronto, ON M5B 1M4 Tel. (416) 363.3171
Telefax: (416)363.59.63

Les Éditions La Liberté Inc.
3020 Chemin Sainte-Foy
Sainte-Foy, PQ G1X 3V6 Tel. (418) 658.3763
Telefax: (418) 658.3763

Federal Publications Inc.
165 University Avenue, Suite 701
Toronto, ON M5H 3B8 Tel. (416) 860.1611
Telefax: (416) 860.1608

Les Publications Fédérales
1185 Université
Montréal, QC H3B 3A7 Tel. (514) 954.1633
Telefax : (514) 954.1635

CHINA – CHINE
China National Publications Import
Export Corporation (CNPIEC)
16 Gongti E. Road, Chaoyang District
P.O. Box 88 or 50
Beijing 100704 PR Tel. (01) 506.6688
Telefax: (01) 506.3101

DENMARK – DANEMARK
Munksgaard Book and Subscription Service
35, Nørre Søgade, P.O. Box 2148
DK-1016 København K Tel. (33) 12.85.70
Telefax: (33) 12.93.87

FINLAND – FINLANDE
Akateeminen Kirjakauppa
Keskuskatu 1, P.O. Box 128
00100 Helsinki
Subscription Services/Agence d'abonnements :
P.O. Box 23
00371 Helsinki Tel. (358 0) 12141
Telefax: (358 0) 121.4450

FRANCE
OECD/OCDE
Mail Orders/Commandes par correspondance:
2, rue André-Pascal
75775 Paris Cedex 16 Tel. (33-1) 45.24.82.00
Telefax: (33-1) 45.24.81.76 or (33-1) 45.24.85.00
Telex: 640048 OCDE

OECD Bookshop/Librairie de l'OCDE :
33, rue Octave-Feuillet
75016 Paris Tel. (33-1) 45.24.81.67
(33-1) 45.24.81.81

Documentation Française
29, quai Voltaire
75007 Paris Tel. 40.15.70.00

Gibert Jeune (Droit-Économie)
6, place Saint-Michel
75006 Paris Tel. 43.25.91.19

Librairie du Commerce International
10, avenue d'Iéna
75016 Paris Tel. 40.73.34.60

Librairie Dunod
Université Paris-Dauphine
Place du Maréchal de Lattre de Tassigny
75016 Paris Tel. (1) 44.05.40.13

Librairie Lavoisier
11, rue Lavoisier
75008 Paris Tel. 42.65.39.95

Librairie L.G.D.J. - Montchrestien
20, rue Soufflot
75005 Paris Tel. 46.33.89.85

Librairie des Sciences Politiques
30, rue Saint-Guillaume
75007 Paris Tel. 45.48.36.02

P.U.F.
49, boulevard Saint-Michel
75005 Paris Tel. 43.25.83.40

Librairie de l'Université
12a, rue Nazareth
13100 Aix-en-Provence Tel. (16) 42.26.18.08

Documentation Française
165, rue Garibaldi
69003 Lyon Tel. (16) 78.63.32.23

Librairie Decitre
29, place Bellecour
69002 Lyon Tel. (16) 72.40.54.54

GERMANY – ALLEMAGNE
OECD Publications and Information Centre
August-Bebel-Allee 6
D-53175 Bonn 2 Tel. (0228) 959.120
Telefax: (0228) 959.12.17

GREECE – GRÈCE
Librairie Kauffmann
Mavrokordatou 9
106 78 Athens Tel. (01) 32.55.321
Telefax: (01) 36.33.967

HONG-KONG
Swindon Book Co. Ltd.
13–15 Lock Road
Kowloon, Hong Kong Tel. 366.80.31
Telefax: 739.49.75

HUNGARY – HONGRIE
Euro Info Service
POB 1271
1464 Budapest Tel. (1) 111.62.16
Telefax : (1) 111.60.61

ICELAND – ISLANDE
Mál Mog Menning
Laugavegi 18, Pósthólf 392
121 Reykjavik Tel. 162.35.23

INDIA – INDE
Oxford Book and Stationery Co.
Scindia House
New Delhi 110001 Tel.(11) 331.5896/5308
Telefax: (11) 332.5993

17 Park Street
Calcutta 700016 Tel. 240832

INDONESIA – INDONÉSIE
Pdii-Lipi
P.O. Box 269/JKSMG/88
Jakarta 12790 Tel. 583467
Telex: 62 875

IRELAND – IRLANDE
TDC Publishers – Library Suppliers
12 North Frederick Street
Dublin 1 Tel. (01) 874.48.35
Telefax: (01) 874.84.16

ISRAEL
Electronic Publications only
Publications électroniques seulement
Sophist Systems Ltd.
71 Allenby Street
Tel-Aviv 65134 Tel. 3-29.00.21
Telefax: 3-29.92.39

ITALY – ITALIE
Libreria Commissionaria Sansoni
Via Duca di Calabria 1/1
50125 Firenze Tel. (055) 64.54.15
Telefax: (055) 64.12.57

Via Bartolini 29
20155 Milano Tel. (02) 36.50.83

Editrice e Libreria Herder
Piazza Montecitorio 120
00186 Roma Tel. 679.46.28
Telefax: 678.47.51

Libreria Hoepli
Via Hoepli 5
20121 Milano Tel. (02) 86.54.46
Telefax: (02) 805.28.86

Libreria Scientifica
Dott. Lucio de Biasio 'Aeiou'
Via Coronelli, 6
20146 Milano Tel. (02) 48.95.45.52
Telefax: (02) 48.95.45.48

JAPAN – JAPON
OECD Publications and Information Centre
Landic Akasaka Building
2-3-4 Akasaka, Minato-ku
Tokyo 107 Tel. (81.3) 3586.2016
Telefax: (81.3) 3584.7929

KOREA – CORÉE
Kyobo Book Centre Co. Ltd.
P.O. Box 1658, Kwang Hwa Moon
Seoul Tel. 730.78.91
Telefax: 735.00.30

MALAYSIA – MALAISIE
Co-operative Bookshop Ltd.
University of Malaya
P.O. Box 1127, Jalan Pantai Baru
59700 Kuala Lumpur
Malaysia Tel. 756.5000/756.5425
Telefax: 757.3661

MEXICO – MEXIQUE
Revistas y Periodicos Internacionales S.A. de C.V.
Florencia 57 - 1004
Mexico, D.F. 06600 Tel. 207.81.00
Telefax : 208.39.79

NETHERLANDS – PAYS-BAS
SDU Uitgeverij Plantijnstraat
Externe Fondsen
Postbus 20014
2500 EA's-Gravenhage Tel. (070) 37.89.880
Voor bestellingen: Telefax: (070) 34.75.778

NEW ZEALAND
NOUVELLE-ZÉLANDE
Legislation Services
P.O. Box 12418
Thorndon, Wellington Tel. (04) 496.5652
Telefax: (04) 496.5698

NORWAY – NORVÈGE
Narvesen Info Center – NIC
Bertrand Narvesens vei 2
P.O. Box 6125 Etterstad
0602 Oslo 6 Tel. (022) 57.33.00
Telefax: (022) 68.19.01

PAKISTAN
Mirza Book Agency
65 Shahrah Quaid-E-Azam
Lahore 54000 Tel. (42) 353.601
Telefax: (42) 231.730

PHILIPPINE – PHILIPPINES
International Book Center
5th Floor, Filipinas Life Bldg.
Ayala Avenue
Metro Manila Tel. 81.96.76
Telex 23312 RHP PH

PORTUGAL
Livraria Portugal
Rua do Carmo 70-74
Apart. 2681
1200 Lisboa Tel.: (01) 347.49.82/5
Telefax: (01) 347.02.64

SINGAPORE – SINGAPOUR
Information Publications Pte. Ltd.
41, Kallang Pudding, No. 04-03
Singapore 1334 Tel. 741.5166
Telefax: 742.9356

SPAIN – ESPAGNE
Mundi-Prensa Libros S.A.
Castelló 37, Apartado 1223
Madrid 28001 Tel. (91) 431.33.99
Telefax: (91) 575.39.98

Libreria Internacional AEDOS
Consejo de Ciento 391
08009 – Barcelona Tel. (93) 488.30.09
Telefax: (93) 487.76.59

Llibreria de la Generalitat
Palau Moja
Rambla dels Estudis, 118
08002 – Barcelona
(Subscripcions) Tel. (93) 318.80.12
(Publicacions) Tel. (93) 302.67.23
Telefax: (93) 412.18.54

SRI LANKA
Centre for Policy Research
c/o Colombo Agencies Ltd.
No. 300-304, Galle Road
Colombo 3 Tel. (1) 574240, 573551-2
Telefax: (1) 575394, 510711

SWEDEN – SUÈDE
Fritzes Information Center
Box 16356
Regeringsgatan 12
106 47 Stockholm Tel. (08) 690.90.90
Telefax: (08) 20.50.21

Subscription Agency/Agence d'abonnements :
Wennergren-Williams Info AB
P.O. Box 1305
171 25 Solna Tel. (08) 705.97.50
Téléfax : (08) 27.00.71

SWITZERLAND – SUISSE
Maditec S.A. (Books and Periodicals - Livres et périodiques)
Chemin des Palettes 4
Case postale 266
1020 Renens Tel. (021) 635.08.65
Telefax: (021) 635.07.80

Librairie Payot S.A.
4, place Pépinet
CP 3212
1002 Lausanne Tel. (021) 341.33.48
Telefax: (021) 341.33.45

Librairie Unilivres
6, rue de Candolle
1205 Genève Tel. (022) 320.26.23
Telefax: (022) 329.73.18

Subscription Agency/Agence d'abonnements :
Dynapresse Marketing S.A.
38 avenue Vibert
1227 Carouge Tel.: (022) 308.07.89
Telefax : (022) 308.07.99

See also – Voir aussi :
OECD Publications and Information Centre
August-Bebel-Allee 6
D-53175 Bonn 2 (Germany) Tel. (0228) 959.120
Telefax: (0228) 959.12.17

TAIWAN – FORMOSE
Good Faith Worldwide Int'l. Co. Ltd.
9th Floor, No. 118, Sec. 2
Chung Hsiao E. Road
Taipei Tel. (02) 391.7396/391.7397
Telefax: (02) 394.9176

THAILAND – THAÏLANDE
Suksit Siam Co. Ltd.
113, 115 Fuang Nakhon Rd.
Opp. Wat Rajbopith
Bangkok 10200 Tel. (662) 225.9531/2
Telefax: (662) 222.5188

TURKEY – TURQUIE
Kültür Yayinlari Is-Türk Ltd. Sti.
Atatürk Bulvari No. 191/Kat 13
Kavaklidere/Ankara Tel. 428.11.40 Ext. 2458
Dolmabahce Cad. No. 29
Besiktas/Istanbul Tel. 260.71.88
Telex: 43482B

UNITED KINGDOM – ROYAUME-UNI
HMSO
Gen. enquiries Tel. (071) 873 0011
Postal orders only:
P.O. Box 276, London SW8 5DT
Personal Callers HMSO Bookshop
49 High Holborn, London WC1V 6HB
Telefax: (071) 873 8200
Branches at: Belfast, Birmingham, Bristol, Edinburgh, Manchester

UNITED STATES – ÉTATS-UNIS
OECD Publications and Information Centre
2001 L Street N.W., Suite 700
Washington, D.C. 20036-4910 Tel. (202) 785.6323
Telefax: (202) 785.0350

VENEZUELA
Libreria del Este
Avda F. Miranda 52, Aptdo. 60337
Edificio Galipán
Caracas 106 Tel. 951.1705/951.2307/951.1297
Telegram: Libreste Caracas

Subscription to OECD periodicals may also be placed through main subscription agencies.

Les abonnements aux publications périodiques de l'OCDE peuvent être souscrits auprès des principales agences d'abonnement.

Orders and inquiries from countries where Distributors have not yet been appointed should be sent to: OECD Publications Service, 2 rue André-Pascal, 75775 Paris Cedex 16, France.

Les commandes provenant de pays où l'OCDE n'a pas encore désigné de distributeur devraient être adressées à : OCDE, Service des Publications, 2, rue André-Pascal, 75775 Paris Cedex 16, France.

12-1993

OECD PUBLICATIONS, 2 rue André-Pascal, 75775 PARIS CEDEX 16
PRINTED IN FRANCE
(22 94 01 1) ISBN 92-64-14057-3 - No. 46585 1994